Blue Planet, Blue God

Blue Planet, Blue God

The Bible and the Sea

Meric Srokosz
and
Rebecca S. Watson

scm press

© Meric Srokosz and Rebecca S. Watson 2017

Published in 2017 by SCM Press
Editorial office
3rd Floor, Invicta House,
108–114 Golden Lane,
London EC1Y 0TG, UK
www.scmpress.co.uk

SCM Press is an imprint of Hymns Ancient & Modern Ltd
(a registered charity)

Hymns Ancient & Modern® is a registered trademark of
Hymns Ancient & Modern Ltd
13A Hellesdon Park Road, Norwich,
Norfolk NR6 5DR, UK.

British Library Cataloguing in Publication data
A catalogue record for this book is available
from the British Library

978 0 334 05633 1

Typeset by Regent Typesetting
Printed and bound by
Ashford Colour Press Ltd.

Contents

Acknowledgements

This book would not have come into being without the support of many people. Our research on 'The Sea in Scripture' was initially funded by the Faraday Institute for Science and Religion, under a Templeton World Charity Foundation grant, and this particular book was then facilitated by a further grant from the Issachar Fund. Without this backing it would never have been written, and we are greatly indebted to them. The director of the Faraday Institute, Professor Bob White, was supportive and encouraged our work and writing throughout. We would like to thank our friends at the Faraday Institute for their input and feedback, among them Ruth Bancewicz, Rachel Simonson, Hilary Marlow, Eleanor Puttock and Shwethaa Raghunathan. Many other members of our families and friends kindly commented on earlier drafts of the manuscript, among them Joyce Chaplin, Nicola McIntyre, Helen James and Tanya Bownds. Stuart Wood was hugely helpful in suggesting relevant poems and other literary extracts relating to our topic at an early stage of writing, and were it not for his input, we might never have developed this aspect of the book. In the last stages of preparation, Joyce Chaplin did an amazing job of proof-reading and uncovered all manner of defects that we had overlooked. Those that remain are, of course, entirely our own.

We are also both grateful for the role played by our families. Rebecca would like to acknowledge that finishing this book has inevitably demanded a lot of time, which has taken her away from her children (if only into the study in the next room) and put a heavier load on her husband, James. She would like to thank James for his support and understanding, and hopes her children will in the end be persuaded that her efforts were worthwhile. Meric would like to thank his wife Anne for her encouragement in completing the book, which took somewhat longer than expected. Various members of his family, friends and church provided encouragement too, and hopefully will enjoy the end results.

Finally, we would like to thank David Shervington, our editor at SCM Press, for advice and encouragement and, of course, for his willingness to publish this book.

All biblical quotations used in the book are from the New Revised Standard Version. The poetry we have quoted is reproduced by the kind permission of the copyright holders. All unacknowledged poems are thought to be in the public domain. We have made every effort to trace copyright holders, and if inadvertently we have failed we would be happy to rectify this in future editions of the book. We acknowledge the following:

The poem 'Water' by Wendell Berry. Copyright © 1971, 2011 by Wendell Berry from *Farming: A Hand Book*. Reprinted by permission of Counterpoint.

J. J. Bola's poem 'Tell them (they have names)'. Copyright © J. J. Bola, and reproduced by permission of the author.

John Masefield's poems 'Cargoes' and 'Dauber' are used by permission of The Society of Authors as the Literary Representative of the Estate of John Masefield.

'The Quaker Graveyard in Nantucket' from *Lord Weary's Castle* by Robert Lowell. Copyright © 1946 by Robert Lowell. Copyright © renewed 1974 by Robert Lowell. Reprinted by permission of Houghton Mifflin Harcourt Publishing Company. All rights reserved.

'Fireflies of the Sea' by James Fenton, from *Yellow Tulips: Poems 1968–2011* by James Fenton (Faber & Faber, 2012). Copyright © James Fenton, 2012. Reproduced by permission of Faber & Faber Ltd.

'Fireflies of the Sea' from *Out of Danger* by James Fenton. Copyright © 1994 by James Fenton. Reprinted by permission of Farrar, Straus and Giroux.

'Fireflies of the Sea' by James Fenton also used by permission of his agents United Agents LLP.

Introduction

... for the earth will be full of the knowledge of [the glory of] the LORD
as the waters cover the sea. (Isaiah 11.9; see also Habakkuk 2.14)

Perhaps you have been puzzled by this verse from Isaiah, and
the similar verse in Habakkuk 2.14? On one level the meaning
is obvious: according to Isaiah 11.9, at some point in the future
people will fully know the presence of God and live in the light
of that knowledge: in justice, peace, righteousness, and holy fear
(cf. 11.2–5), so that there is harmony between species (cf. Isaiah
65.17–25).[1] In Habakkuk 2.14, it is the glory of God – his splendour, power and majesty, but also his awesome presence – that
will be acknowledged throughout the earth. But on another level
it contains a mystery: what does it mean for the waters to cover
the sea? Surely the waters *are* the sea?

We shall return to the meaning of these verses later. For now it
serves as a starting point for outlining what we are attempting to
do in this book. The aim is to examine what the Bible says about
the sea, hence the 'Blue Planet' of the title. We have found that
the passages concerned with the sea challenge our thinking about
God's relationship with this important part of his creation and
how he might feel about it, hence the 'Blue God' part of the title.
Much of what the Bible says regarding the sea seems mysterious
at first sight, but this forces us to reflect more carefully on its
message. Even more, that very process will challenge our thinking
on many issues that we as human beings face in today's world.

Little has been written on the Bible's view of the sea. Traditional
scholarship has tended to see the Israelites of the Old Testament
as somewhat divorced from maritime concerns (with the possible
exception of the story of Jonah). However, as we hope to show
in what follows, this approach neglects the large amount of Old

Testament material on the sea. New Testament scholarship has paid a little more attention to this topic, not least because Jesus recruited fishermen to be his disciples and spent some time in boats on the Sea of Galilee. In addition, Paul's missionary journeys involved Mediterranean voyages and both Jesus and Paul experienced storms at sea. Despite this, there has been nothing written that looks at both the Old and New Testament material on the sea as a whole. This book attempts to remedy that lack to some degree.

The book came about because one of us (Meric) is an oceanographer and goes to sea to carry out research. In reading the Bible he became aware of the wealth of material on the sea but found that biblical scholars had written little on this subject. He enlisted the help of a biblical scholar (Rebecca) and initiated the 'Sea in Scripture' project at the Faraday Institute for Science and Religion (where he was the Associate Director at the time). The outcome of this research is to be twofold: both this book and a more academic monograph. The book has resulted from our growing conviction as we studied the topic that the biblical material on the sea provides a 'lens' through which we may be challenged about our attitudes and behaviour – essentially, how we are to live in what the Bible understands to be God's world.

The Bible is the product of a faith community and written for people of faith. As such, it seeks to elicit a response and to effect changes in people's lives. If study of the Bible evokes in us a desire to respond to God and to live our lives in the light of his character and purposes both for us and for his creation then it will have fulfilled its purpose. It is our conviction, though, that it can offer provocation and challenge, vision and hope, to anyone, regardless of their personal beliefs. Its call to justice and integrity, love and responsibility in our conduct towards others and to the rest of creation (among other virtues) can speak to all of us. In addition, as an ancient text, even the most recent parts of which are not much less than 2,000 years old, it offers a window on to a way of living and perceiving the world that is far removed from the modern Western, individualistic and largely urban perspective that we inhabit. This cultural gap can enable us to re-evaluate some of the often unreflective assumptions that motivate our responses, not least in relation to some of the most pressing modern issues that demand our attention; these will be considered in what follows. Our hope is that readers will be provoked and challenged by what follows and then respond accordingly. If that happens, then the book will have achieved its aim.

Note

1 The converse of the idea that peace in creation results from righteousness is reflected in Leviticus 26.22, in the notion that human disobedience will result in attacks from wild animals.

I

The Sea and Salvation

When we tell people what we are working on – that is, what the Bible says about the sea[1] – it often elicits the same reaction: 'The people of the Bible didn't like the sea, did they? They were afraid of it.' It is true that there are indeed some passages in the Bible that portray the sea as terrible and threatening, but this is only one facet of a much more diverse picture, hence this book. In particular, at the very beginning of the Bible (Genesis 1.10) God gathers the waters and calls them seas, and 'God saw that it was good'. He then (Genesis 1.21) populates the ocean with creatures and again sees that it was good. Our conviction, as expressed in this book, is that the sea is a good part of God's creation and what the Bible says about it has relevance to us today.

Interestingly, in the Gospel accounts we find that a large part of Jesus' ministry, especially the early part, was centred on the Sea of Galilee and surrounding districts (as in Mark 1–8). Additionally, at least one post-resurrection encounter between Jesus and his disciples is described as occurring by the Sea of Galilee (John 21.1–14).[2] It is also well known that several of the disciples chosen by Jesus to follow him were fishermen, who were henceforth called to fish for people (Matthew 4.18–22). So it is apparent that Jesus was familiar with the sea and the creatures in it as part of his daily life,[3] a fact that is often overlooked or unacknowledged and which this book seeks to re-emphasize.

We begin by looking in this chapter at one of the major themes in the Bible – salvation – and how that is linked to stories of the sea. Being alone and vulnerable on the wide sea can create a sense of mortality and dependence on God that is seldom replicated elsewhere. In some cases, this can even provoke a total reprioritization and change of life. However, in the Bible, salvation at sea also often marks a new beginning, as we shall see.

The sea and salvation

If you were asked to name stories from the Bible involving the sea, the chances are that the top five to come to mind would be Noah's Flood (which, strictly speaking, is not really about the sea),[4] the crossing of the Red Sea during the Exodus, Jonah and the 'big fish', and Jesus' miracles of stilling the storm and of walking on water. The first thing to note is that, in all of these stories, the main protagonists voluntarily take to the sea. For Noah, the Ark was the only safe place to be in the oncoming Deluge (Genesis 6—9), and it became an early symbol of the Church – a safe haven, holding God's chosen people and protecting them from the chaotic world outside. Jonah somehow imagined he could escape God's call by being shipped to the other end of the known world. Of course, he could not – but he certainly did not dismiss the idea of taking to the sea, and viewed it as a way of escape (Jonah 1—2). Amazingly, the Israelites fleeing Egypt headed straight for the Red Sea (Exodus 14), only to experience the key act of salvation that formed them as God's people and which became a fundamental aspect of their national story and religious identity. It is the event that remained at the heart of their faith and hopes for the future. For Jesus and his disciples, the Sea of Galilee was a constant scene of their activity, and crossing the lake by boat (or on foot!) was an easier, and more obvious, way to travel than walking round the shore (Mark 4.35–41, 6.45–52 and parallels in other Gospels).

Nonetheless, this is not to say that the sea is always a safe place to be in these stories: salvation from the sea is a common theme throughout these accounts. In the case of the story of the Flood, strictly speaking (and actually quite importantly), although the waters covered the earth and hence were quite sea-like, the sea is not mentioned. Water poured out through the windows of heaven (as rain) and the fountains of the deep were opened so that water welled up from beneath (like rivers and streams). In other words, its causes were like those of other floods experienced in the ancient world – excess rain and overcharged rivers – but now on an unimaginable scale. What is made clear is that the Flood was a divinely instigated event, prompted by the wickedness of humanity, or the corruption of 'all flesh':

> ⁵The LORD saw that the wickedness of humankind was great in the earth, and that every inclination of the thoughts of their hearts was only evil continually. ⁶And the LORD was sorry that he had made humankind on the earth, and it grieved him to his heart. ⁷So the LORD said, 'I will blot out from the earth the human beings I have created – people together with animals and creeping things and birds of the air, for I am sorry that I have made them.' (Genesis 6.5–7)

> ¹¹Now the earth was corrupt in God's sight, and the earth was filled with violence. ¹²And God saw that the earth was corrupt; for all flesh had corrupted its ways upon the earth. ¹³And God said to Noah, 'I have determined to make an end of all flesh, for the earth is filled with violence because of them; now I am going to destroy them along with the earth.' (Genesis 6.11–13)

The Ark, however, was the means of saving the chosen few – righteous Noah and his family – together with representatives of all non-aquatic species, in order that it might cradle life through the Flood and provide a basis for a new beginning.

The saving of a chosen people (though now more numerous) is also the basis of the Exodus story, as also is the idea of a new beginning. This is the moment when the people cross out of Egypt, escape their Egyptian masters for good, and are forged together under God as his people, and he as their God. The sea is obedient to God – being piled up or dried (depending on the version you read) in order to let the people pass through, and then overwhelming the Egyptian enemy (Exodus 14.13–31).

In the case of Jonah, this is a story of the wouldn't-be prophet being saved from himself, in order that the Ninevites might be saved. The book of Jonah is full of parody and surprise, constantly reversing stereotypes: the Israelite prophet, who does not understand or respond to God, is sent to warn the inhabitants of a wicked foreign city to expect divine judgement for their sins, and finds (contrary to his expectations, and perhaps ours too) that they, unlike him, respond with faith. He knows God is concerned for Nineveh, but somehow imagines that he can flee from him by leaving Israelite soil. He confesses to the mariners that he worships 'the God of heaven, who made the sea and the dry land' – and is apparently unaware of the irony of this, whereas their much more perceptive response is to ask fearfully, 'What is this that you have done?' He had expected to be safe from God's call on board ship, and indeed he even slept through the storm till the captain woke him, despite the fact that everyone else was crying out to their gods. However, when he is cast into the water, when we imagine all will be lost, that is when salvation comes, in the form of the 'big fish'. Of course, it not only saves Jonah from drowning, but miraculously transports him back to dry land.

The theme of salvation is very clear in respect of the stilling of the storm (Mark 4.35–41). Here, the people crying out for help are the core group who will form another new start for humanity – not the few chosen ones surviving the Deluge (as in the case of Noah and his family), nor the chosen people of Israel (as with the Red Sea crossing), but Jesus' disciples.[5] These were the people who would be the foundation for the Church and a new era of God's engagement with a differently constituted people (Acts 1–2). In the story of Jesus walking on water (Mark 6.45–52), the ability to stride across the waves is the feature we tend to latch on to, but actually the setting is a stormy sea, and the disciples' fear is associated with their peril, not simply their awe and alarm at Jesus' appearance. Again, he stills the storm.

In each case, then, salvation occurs at sea. However, at the same time something important happens at sea or on the sea,[6] and a new beginning is made. Noah was saved not from the sea, of course, but from destruction and the eradication of most of creation as a result of human sin. The outcome was a fresh start for creation and a new relationship with God.[7] At the Red Sea, the water was an instrument of God's judgement again, as the waves destroyed a hostile army led by Pharaoh, who had caused suffering to the chosen people and attempted to thwart God's purposes. This enabled a new beginning, the transition from slavery to freedom as the people of Israel journeyed to the Promised Land. For Jonah, in the sea he lost his sense of self-determination and independence from God, and learnt about the scope of divine power, and about obedience. When he emerged from the fish, he accepted his call and went to Nineveh to deliver the message with which he had been commissioned. The disciples understood on the sea that God alone could control the wind and waves: it was a moment, then, that revealed Jesus' identity and that addressed their lack of faith. In each case, a 'sea-change' is effected.

However, the story of danger and new beginnings through water (if not the sea) is not only confined to the Bible. It came to be expressed in due course on an individual basis through baptism: the believer is understood to pass through the waters of death, dying to sin and being born again to new life in Christ (Romans 6.1–4). Here salvation is spiritual (though ultimately physical at the resurrection), and thus is distinct from the judgement enacted against corruption on the earth through the floodwaters of the Deluge, or from the destruction of evil powers at the Red Sea,[8] or from the stilling of physical waves at the Sea of Galilee.

It is a marked feature of modern life that we are reluctant to pass moral judgement, to talk about sin or to think of being saved from evil, and we tend to find the idea of judgement problematic. However, if we return to the biblical references to the sea, these help us to understand that the objects of our fear or the problems that we need to address are not confined to 'sin' in a purely theological sense, nor are they always embodied in other people, or indeed in ourselves – though they often *are* found here, at least in part. The sea in the passages we have looked at represents life-threatening danger, but not necessarily moral danger.[9] Rather, the stories concern fear followed by salvation from the source of that fear. The Red Sea crossing is associated with escape from the terrifying threat of the pursuing Egyptian army, whereas Jesus' stilling of the storm or walking on water represents his power over the natural forces of wind, sea and storm, despite the panic they induced in the helpless disciples. For Jonah, taking to the sea reflected his fear (or abhorrence) of his mission and desire to escape God's call. By reading the biblical passages relating to the sea, we too can gain new insights into many of our deepest worries and fears.

This book

References to the sea are not merely confined to questions of fear and salvation, significant as these issues are. This book explores a wide variety of themes that emerge from the study of the sea in the Bible, ranging from the impact of the sea on spirituality, to the ocean as a sacred space itself; from the relation of God to his creation as seen in respect of the marine environment, to our own, human, impact on its manifold life. As the sea has for millennia been a crucial vehicle for trade and migration, biblical allusions offer much material for reflection on economics, trade and the movement of human populations. However, one of the most profound contributions of biblical thinking on the sea is metaphorical, for the sea can represent the unleashing of great chaotic forces beyond our control. Biblical evocations of the threat of the sea destabilizing the very world order speak powerfully to modern fears of the chaotic: terrorism, pandemic, climate change, nuclear disaster, and individual or collective annihilation.

This book, then, touches on some of the most fundamental issues for our time, such as economics, migration and climate change, but it also offers perspectives on some of the most enduring questions for humanity: those of meaning and purpose, of our place in the world, and the need to allay our fears and seek stability despite threats to the status quo. Biblical passages concerning the sea provide challenge and provocation in our thinking about our relationship with God, with creation and with the creatures in it. The following chapters focus on the most important themes that have emerged from our study of the Bible:

- The profound effect the sea may have on spirituality (Chapter 2).
- The relation of God and humanity to the life of the sea (Chapters 3 and 4).
- The sea as a space that has a sacred quality (Chapter 5).
- The sea as 'chaotic' and dangerous, but also vulnerable and in need of protection (Chapters 6 and 7).
- Economics, trade and travel on the sea (Chapter 8).

The chapters address different aspects of the Bible's view of the sea, but in each case we link the biblical material to current issues that directly or indirectly affect all of our lives. We seek to bring biblical insights to bear on contemporary problems, with results that are at times surprising and thought-provoking. We hope that this will evoke in the reader the same response as it does in us: how should we then live?[10] The final chapter, 'Blue Planet, Blue God', seeks to bring together the themes and understanding gained to give a holistic picture.

Each chapter begins with a brief look at a relevant biblical passage, then draws on the current scientific understanding of the ocean, before returning to examine the relevant biblical material in more detail. The book does not aim to be comprehensive; rather, we have selected aspects of the Bible's perspectives on the sea that we think are interesting and less well known and that are likely to stimulate thought and action. To aid in this, each of the main chapters ends with a Key Message, a Challenge, some questions for Reflection and Discussion, followed by some thoughts on Action that could result. We want you, the reader, not just to find the book intellectually stimulating (though hopefully it is), but also to respond to its impetus to live life more in line with what we understand to be God's intentions for his people and his creation.

In this chapter, we started with the idea that the sea can represent what we fear, and emergence from it represents salvation. It is hoped that the following chapters will take you on a journey to confront some of today's pressing concerns, and that our route will help you navigate a way through. However, we also want to leave you with a sense of delight in the sea, and a conviction that it is an important and valuable part of God's creation. We hope that this journey will evoke in you, the reader, the same response as it does in us: a desire to preserve and treasure this wonderful, fearsome and vital part of creation, but also a wish to allow ourselves to be challenged by biblical perspectives on the sea. These perspectives, we believe, connect with many other aspects of our lives, and invite us, above all, to live in the light of the sea. Perhaps it would be appropriate to begin with the prayer of St Brendan the Navigator[11] as we embark on this exploration of the sea in the Bible:

Shall I abandon, O King of mysteries, the soft comforts of
  home?
Shall I turn my back on my native land, and turn my face
  towards the sea?
Shall I put myself wholly at your mercy,
without silver, without a horse,
without fame, without honour?
Shall I throw myself wholly upon You,
without sword and shield, without food and drink,
without a bed to lie on?
Shall I say farewell to my beautiful land, placing myself under
  Your yoke?
Shall I pour out my heart to You, confessing my manifold sins
  and begging forgiveness,
tears streaming down my cheeks?
Shall I leave the prints of my knees on the sandy beach,
a record of my final prayer in my native land?
Shall I then suffer every kind of wound that the sea can inflict?
Shall I take my tiny boat across the wide sparkling ocean?
O King of the Glorious Heaven, shall I go of my own choice
  upon the sea?
O Christ, will You help me on the wild waves?

Notes

1 This book arises from a project at the Faraday Institute for Science
and Religion, entitled 'The Sea in Scripture'.

2 Also known as the Sea of Tiberias or Lake of Gennesaret.

3 He is even portrayed as looking to the sea to meet his needs in the
provision of the money for the temple tax through a coin in the mouth of
a fish (Matthew 17.24–27).

4 There is no mention of the sea in Genesis 6—9, except once in relation
to the 'fish of the sea' (9.2). Instead, the Hebrew uses a very specific word,
hammabbûl, which is translated 'the flood' in English versions of the Bible.
This term, with just one exception, only occurs in relation to the deluge of
Noah's time, even though there are many other more common expressions
that might have been used to describe rain, flooding, rivers, and the sea.
In order to understand its meaning better, we need to look at the other

occurrence of this word, in Psalm 29.10, where God is described as being 'enthroned over *hammabbûl*'. In this instance, it is clear that the term describes the waters held in heaven, which were understood as the source of the rain. (There was also thought to be a related water source under the temple, which fed the streams and other watercourses from beneath.)

When we look back at the story in Genesis 6—9 more carefully, we find that the Hebrew refers not to 'a *mabbûl*' but to 'the *mabbûl*', suggesting that a particular object is in mind, not just an event or usual weather phenomena. Second, the 'flood' is something 'brought on the earth', suggesting that it previously existed prior to this time, but at some remove from the earth and then brought down upon it. Both of these considerations fit with the meaning of the word in Psalm 29, as a celestial reservoir that was 'brought down' to flood the earth. If we look more closely at Genesis 7, we find that although it mentions rain (vv. 4, 12), something more fundamental is happening than pluvial flooding. According to v. 11, 'on that day all the fountains of the great deep burst forth, and the windows of the heavens were opened'. In other words, the 'deep' beneath the earth was released on to it to flood it from below, and water was likewise disgorged from heaven above. Scholars have noted that this reverses the separation of the waters above and below the earth that was described as happening on the second day of creation. This is not just a 'flood' but, like the destruction of all life on earth, reverses creation itself.

5 Note too that the disciples were chosen by Jesus (Mark 3.13–15), just as Noah and his family and Israel had been chosen by God.

6 Of course, Jesus and the disciples were on the Sea of Galilee, technically a very large lake (Luke uses the term for a lake in his Gospel). Nevertheless, in a small boat, a storm on a lake can be just as terrifying as being in a storm on the open ocean in a somewhat larger ship, as in the case of Jonah on the Mediterranean.

7 After the Flood, God institutes a new covenant with humans and with the wider creation (Genesis 9.1–17; note the echoes of the Genesis 1 creation story).

8 Though note that Paul in 1 Corinthians 10.1–2 uses the passing through the Red Sea as a metaphor for baptism.

9 Certainly, the Flood is the means of destroying a corrupt world, but it is in many ways the exception that proves the rule. Unlike the sea in the other stories considered here, it is a uniquely generated instrument of divine punishment for a specific purpose – one that God solemnly promises never to repeat (Genesis 9.8–17). Nor is the Flood to be confused with the sea itself, which has its rightful place in a good and well-ordered creation.

10 To plagiarize the title of one of Francis Schaeffer's books.

11 St Brendan (c. 484–c. 587) was an Irish monk, a traveller and a founder of monasteries. On his first voyage he travelled to islands off Scotland and to Wales and Brittany, founding monasteries on the way. He is

famous for his second voyage, which lasted seven years and is described in the ninth-century *Voyage of St Brendan (Navigatio Sancti Brendani)*. He claimed to have discovered a new island in the far west. It is thought he may have gone as far as Iceland, Greenland and even North America, well ahead of the Vikings in the tenth century and Christopher Columbus in 1492. Various versions of the story may be found in W. R. J. Barron and G. S. Burgess (eds), 2005, *The Voyage of St Brendan: Representative Versions of the Legend in English Translation* (Exeter: University of Exeter Press). The possibility that St Brendan reached North America was tested successfully by Tim Severin, who built a boat similar to that used by St Brendan, and sailed it with a small crew from Ireland to North America, via the Faeroes, Iceland and Greenland (see T. Severin, 1978, *The Brendan Voyage: Across the Atlantic in a Leather Boat*, London: Hutchinson & Co.).

2

The Sea and Spirituality

God of the storm

Close encounters with the sea can have a profound effect on a person's spirituality, influencing their understanding of the world, themselves and God, and of the relationship between all three. Probably most impactful is the sensation of feeling out of place on the waves, particularly in the experience of being caught out in a storm. In such a situation, seafarers are powerfully impressed by a sense of helplessness, of knowing what it is to be dependent on forces beyond themselves and (in many cases) on God. Within a biblical context, this is described in Psalm 107.23–32:

> ²³Some went down to the sea in ships,
>   doing business on the mighty waters;
> ²⁴they saw the deeds of the LORD,
>   his wondrous works in the deep.
> ²⁵For he commanded and raised the stormy wind,
>   which lifted up the waves of the sea.
> ²⁶They mounted up to heaven, they went down to the depths;
>   their courage melted away in their calamity;
> ²⁷they reeled and staggered like drunkards,
>   and were at their wits' end.
> ²⁸Then they cried to the LORD in their trouble,
>   and he brought them out from their distress;
> ²⁹he made the storm be still,
>   and the waves of the sea were hushed.
> ³⁰Then they were glad because they had quiet,
>   and he brought them to their desired haven.

[31]Let them thank the LORD for his steadfast love,
    for his wonderful works to humankind.
[32]Let them extol him in the congregation of the people,
    and praise him in the assembly of the elders.

(Psalm 107.23–32)

The starting point of going down to sea in ships, 'doing business' or 'working' on the great waters, reflects the preoccupations of everyday human commercial activity. However, once caught in a storm, these people know their dependence on God and 'cry to the LORD in their trouble' and deep distress. Interestingly, the biblical author understands God as responsible both for the storm and for delivering the sailors by calming the very waves that he had earlier lifted up. The storm itself is not depicted in a negative way. It is not a punishment or a sign of divine anger, but is understood as one of God's 'wondrous works'. This is the same terminology as is applied to his great deeds such as the deliverance of the people of Israel from Egypt at the Red Sea: it is a sign of his power and a reason for wonder and awe. The stirring up of a storm is an aspect of God's mighty work, even though it causes terror and danger to the sailors. Thankfully, he also responds to prayer and quietens the storm for the sailors' relief!

'O hear us as we cry to thee'

Despite the millennia and technological advances that separate us from the biblical authors, even now fear of the sea has deep roots within our own culture. The naval hymn, 'Eternal Father, strong to save', with its refrain 'O hear us as we cry to thee, For those in peril on the sea', powerfully encapsulates this, as do the evocations of life-threatening storms in the great nautical novels of Herman Melville, Joseph Conrad, Robert Louis Stevenson and others. John Masefield provides a particularly striking window into the intense suffering and danger experienced by sailors by depicting the extreme risk and privation that counteracts the compelling draw of the sea, the 'call that may not be denied'. Here is

an extract from his portrayal of the attempt to round Cape Horn
in his 1912 poem, *Dauber*:

Darkness came down—half darkness—in a whirl;
The sky went out, the waters disappeared.
He felt a shocking pressure of blowing hurl
The ship upon her side. The darkness speared
At her with wind; she staggered, she careered,
Then down she lay. The Dauber felt her go;
He saw his yard tilt downwards. Then the snow

Whirled all about—dense, multitudinous, cold.
Mixed with the wind's one devilish thrust and shriek,
Which whiffled out men's tears, deafened, took hold,
Flattening the flying drift against the cheek.
The yards buckled and bent, man could not speak.
The ship lay on her broadside; the wind's sound
Had devilish malice at having got her downed.

How long the gale had blown he could not tell,
Only the world had changed, his life had died.
A moment now was everlasting hell.
Nature an onslaught from the weather side,
A withering rush of death, a frost that cried,
Shrieked, till he withered at the heart; a hail
Plastered his oilskins with an icy mail.

'Cut!' yelled his mate. He looked—the sail was gone,
Blown into rags in the first furious squall;
The tatters drummed the devil's tattoo. On
The buckling yard a block thumped like a mall.
The ship lay—the sea smote her, the wind's bawl
Came, 'loo, loo, loo!' The devil cried his hounds
On to the poor spent stag strayed in his bounds.

'Cut! Ease her!' yelled his mate; the Dauber heard.
His mate wormed up the tilted yard and slashed,
A rag of canvas skimmed like a darting bird.

The snow whirled, the ship bowed to it, the gear lashed,
The sea-tops were cut off and flung down smashed;
Tatters of shouts were flung, the rags of yells—
And clang, clang, clang, below beat the two bells.

'O God!' the Dauber moaned. A roaring rang,
Blasting the royals like a cannonade;
The backstays parted with a crackling clang,
The upper spars were snapped like twigs decayed—
Snapped at their heels, their jagged splinters splayed,
Like white and ghastly hairs erect with fear.
The Mate yelled, 'Gone, by God, and pitched them clear!'

'Up!' yelled the Bosun; 'up and clear the wreck!'
The Dauber followed where he led: below
He caught one giddy glimpsing of the deck
Filled with white water, as though heaped with snow.
He saw the streamers of the rigging blow
Straight out like pennons from the splintered mast,
Then, all sense dimmed, all was an icy blast

Roaring from nether hell and filled with ice,
Roaring and crashing on the jerking stage,
An utter bridle given to utter vice,
Limitless power mad with endless rage
Withering the soul; a minute seemed an age.
He clutched and hacked at ropes, at rags of sail.
Thinking that comfort was a fairy-tale.

Told long ago—long, long ago—long since
Heard of in other lives—imagined, dreamed—
There where the basest beggar was a prince
To him in torment where the tempest screamed,
Comfort and warmth and ease no longer seemed
Things that a man could know: soul, body, brain,
Knew nothing but the wind, the cold, the pain.

Experiencing God in the storm

The prospect of death at sea, whether imminent or prospective, has an almost unique capacity to enable vulnerable sailors to recognize what is most salient in life. John Wesley, on a ship crossing the Atlantic en route to Georgia, records:

> At eleven at night I was waked by a great noise. I soon found there was no danger. But the bare apprehension of it gave me a lively conviction what manner of men those ought to be who are every moment on the brink of eternity. (Friday, 31 October 1735)[1]

Over the succeeding weeks, his fear of death troubled him, and he self-consciously drew on the pattern of Jesus' disciples' fear (Mark 4.40) in asking himself, '"How is it that thou hast no faith? being still unwilling to die"' (Friday, 23 January 1736). As a result, he found himself being drawn to join in worship with the German Moravians on board, since he longed to be able to emulate their faith and fearlessness in the face of life-threatening storms, which contrasted sharply with his own fear and that of his compatriots.

Wesley's experience of the sea as providing a context for his faith to be tested, a moment of truth in which he would come to a realization that could otherwise have eluded him, is far from unique. It is well known that John Newton, the Anglican priest and author of the hymn 'Amazing Grace', had turned, according to his own account, from a life of sin to grace after escaping death in a prolonged ordeal at sea. This began with his ship nearly sinking in a storm and having to be constantly plugged and bailed, and it then continued through the ordeal of attempting to reach safety despite severe food shortages, a mistaken sighting of land, becalming and very nearly running out of water. Parallel to the extreme physical challenges that attended this experience, Newton seems to have begun to undergo a no less strenuous spiritual transformation. After surviving the initial onslaught of the storm, Newton's mind began to turn to the faith that he had so vocally rejected, but was initially convinced that 'there never was, nor could be, such a sinner as myself, ... I concluded, at first, that my sins were too great to be forgiven.'[2]

This oppressive sense of guilt was not helped by the captain's incessant joking that Newton must have been the cause of their troubles and should be cast overboard like Jonah in order to save the others from death. Newton recalls:

> He did not intend to make the experiment, but the continual repetition of this in my ears, gave me much uneasiness, especially as my conscience seconded his words, I thought it very probable, that all that had befallen us was on my account. I was, at last, found out by the powerful hand of God, and condemned in my own breast.

However, a progressive series of more favourable happenings, through which the mariners eluded death, persuaded Newton that

> I thought I saw the hand of God displayed in our favour; I began to pray – I could not utter the prayer of faith; I could not draw near to a reconciled God; and call him Father. My prayer was like the cry of the ravens, which yet the Lord does not disdain to hear. I now began to think of that Jesus, whom I had so often derided; I recollected the particulars of his life, and of his death; a death for sins not his own, but, as I remembered, for the sake of those, who, in their distress, should put their trust in him. And now I chiefly wanted evidence.

Newton began to turn to the Scriptures and to prayer whenever opportunity afforded, with hope and fear alternately being excited by each turn of events. Gradually, however,

> I began to conceive hopes greater than all my fears especially when, at the time we were ready to give up all for lost, and despair was taking place in every countenance, I saw the wind come about to the very point we wished it, so as best to suit that broken part of the ship which must be kept out of the water, and to blow so gentle as our few remaining sails could bear; and thus it continued, without any observable alteration or increase, though at an unsettled time of the year, till we once more were

called up to see the land, and were convinced that it was land indeed.

Once they made landfall, the last of their meagre food supply boiling in a pot, a storm at once arose that would have ruined the ship and drowned its weakened occupants had they not first reached the safety of the harbour. 'About this time', said Newton, 'I began to know that there is a God that hears and answers prayer.' His faith journey continued once on land and was clearly fed, even in the dire conditions on the ship, by prayer and Bible study. Undoubtedly, however, the interplay of life-threatening danger and the prospect of faith as the only framework that could offer an explanation or hope in this desperate predicament provided the conditions that enabled this to happen: the religious perspective that he had brutally mocked suddenly impinged with intense urgency on his consciousness as the only means to understand his plight and deliverance, to seek help, or to prepare for what might happen next. Being baptized is symbolic of passing through the waters of death and being born again as a means of expressing the spiritual transformation undergone by the believer (Romans 6.4). For Newton, this transition from death to life was a process that he underwent literally, in experiencing deliverance from what seemed like certain death at sea; his spiritual journey back to faith followed in the wake of this profound embodied experience.

Awe and wonder

The power of the sea to raise consciousness of the transcendent and to prompt the observer to ask ultimate questions is not, however, confined to situations of dire need. The sheer scale and beauty of the sea, despite the fragility of those who sail on it, has always been a stimulus to awe. Wesley himself was struck, on viewing the Needles, by how

> the ragged rocks, with the waves dashing and foaming at the foot of them, and the white side of the island rising to such a

height, perpendicular from the beach, gave a strong idea of Him
that spanneth the heavens, and holdeth the waters in the hollow
of His hand! (Wednesday, 10 December 1735)

Nonetheless, the sense of perspective afforded by the sea is also
accessible, in some measure, from land. Recall Keats's invitation
(from 'On the Sea'):

> Oh ye! Who have your eye-balls vexed and tired,
> Feast them upon the wideness of the Sea;
>   Oh ye! Whose ears are dimmed with uproar rude,
> Or fed too much with cloying melody, –
>   Sit ye near some old cavern's mouth, and brood.

Of itself, the sea invites contemplation and a wider perspective
than usually enters our consciousness. This sense of perspective,
transcending anything that may be experienced on land, and
inviting a sense of being part of an unimaginably vast creation, is
one of the greatest gifts the sea can bring. Walt Whitman summed
this up in the poem 'On the Beach at Night Alone' in words incorporated into Vaughan Williams's Sea Symphony:

> On the beach at night alone,
> As the old mother sways her to and fro singing her husky song,
> As I watch the bright stars shining, I think a thought of the clef
>   of the universes and of the future.
> A vast similitude interlocks all,
> All spheres, grown, ungrown, small, large, suns, moons,
>   planets,
> All distances of place however wide,
> All distances of time, all inanimate forms,
> All souls, all living bodies though they be ever so different, or
>   in different worlds,
> All gaseous, watery, vegetable, mineral processes, the fishes,
>   the brutes,
> All nations, colors, barbarisms, civilizations, languages,
> All identities that have existed or may exist on this globe, or
>   any globe,

All lives and deaths, all of the past, present, future,
This vast similitude spans them, and always has spann'd,
And shall forever span them and compactly hold and enclose
  them.

The sea invites questions of our place in the world, our relation
to God and to other species. Our insignificance in the scale of
space and time is impressed upon our consciousness by the vastness of the ocean. Encounters with the sea (whether literal or
literary) can provoke a sense of transcendence and connectedness
in the midst of the blinding rush of modern life. It is not surprising that the ceaseless, unpredictable and even violent energy and
movement of the sea, its unfathomable depth and vastness, its
beauty and ferocity, can provoke an awareness of powers and
purposes beyond ourselves, and even a sense of being a frail and
transient part of something much greater than humanity can be. It
provokes questions about the nature of the universe, the presence
or absence of benevolent or hostile forces within it, and our place
within it. Even in a conceptual world preoccupied with material
reality and fragmentizing analysis, this endlessly moving and
powerful force that occupies 71% of the earth's surface perhaps
uniquely invites a perspective that attempts to encompass the
totality of the Earth, and tugs our emotional and sensory perceptions towards questions of transcendence. This particular response
to the sea, perhaps above all others, is attested across time and
space, despite differences in how such spiritual impulses are out-
worked in various contexts.

However, the experience of awe and wonder at the beauty and
power of the sea, and a sense of human finitude against its scale
and ferocity, is not confined to a specifically religious mindset.
Writing from a secular viewpoint, Ellen MacArthur[3] conjures up
the extraordinary qualities of the sea,[4]

> the mesmerizing beauty of the ferocious waves, foaming crests
> and seas so large and rolling you could have built villages in their
> valleys ... seeing a pod of dolphins jump out of a mountainside
> of a seawave, seeing that for them it was probably just another
> day at the office.

She appreciates that 'the power of nature out here knows no bounds – we are so lucky to live on an earth so full of treasure' but at the same time perceives exquisite beauty in little things too. While attempting to repair storm damage, suddenly being struck by

> a mesmerizing flow of bright blue water flying beneath her hull ... I smiled out loud (if you can do that!) and bent down to touch this beautiful miracle of life. Just water – but in that second it was priceless.

She likewise identifies aspects of living at sea which are lost on land,

> the connection that you have with life itself, the rhythms of the natural world, and that feeling that you are seeing something special. Sailing at night is like peeking through a forgotten window to a timeless world beyond, which you have temporarily misplaced. There is no reason on earth that we cannot still see this beauty on land, but due to so many distractions we so rarely do.

An important part of this experience, complementing feelings of awe, beauty and transcendence, is an awareness of her own insignificance. Caught in a gale, 'with winds gusting over 50 to 55 in the squalls', she logged,

> It's about to go dark down here, and the waves are no smaller ... In fact, now we've gybed they seem bigger and more powerful than before. I am completely in awe of this place. The beauty of the immense rolling waves is endless and there is a kind of eternal feeling about their majestic rolling that will live on forever, just watching them roll along – with nothing to stop them makes Mobi [the boat] and I feel completely insignificant – they are hardly aware of our tiny presence on their surface ... I stand in the cockpit and stare – I think I must be the luckiest person in the world to be here seeing, feeling, smelling and touching all this with my own eyes and senses – I feel alive ...

Though it is quite frightening being here and feeling poor Mobi being literally hurled down the waves as she was earlier ... I am glad we have come down here and seen this storm ... It's a reminder of how small and insignificant we are on this planet – but at the same time what a responsibility we have towards its protection.

A little later, she records: 'As I pumped out I occasionally glanced up at the total turmoil of the ocean's surface and marvelled at the raw power of the wilderness for thousands of miles around me.' Conversely, a calm, flat sea brought an equal sense of 'feeling utterly helpless and unable to make a difference'.

A sense of higher powers enabling her passage also played a part in the struggle for survival and success. MacArthur scribbled down in her log something her father said to her: 'at the end of the day the stars will either be with you or against you ... and I think they might be with you'. Later,

as we crossed the Equator, we were over one day ahead of Francis [the previous record holder], and I called out to the heavens in delight. I had thought long and hard about what gift I would give to Neptune as we crossed the line, and had settled on a tiny silver charm I'd had round my neck. I wanted to give him something that mattered to show our appreciation for allowing us to pass there again ... I was quite emotional as I broke open my tiny bottle of champagne and flung the charm into the sea. Grateful emotion though.

Moments of intense gratitude bordering on the religious also feature in her journey. When she finally managed to induce a broken generator to run at a normal temperature, 'I sank to my knees with happiness'. The sense of 'allowing us to pass there' chimes in with a passage in Ezekiel 28 that we shall be returning to later. Here the author of Ezekiel expresses his horror of the hubristic confidence of the seafaring traders from ancient Tyre: like MacArthur, he instinctively recognized that we have no right of passage here, but are insignificant and dependent on God or other (transcendent) forces outside our control or understanding.

For Newton, the sea facilitated a conversion experience and a profound sense of God's 'amazing grace, who saved a wretch like me'; for Wesley, already a man of deep piety, the sea forced him to discover the limits of his faith, since when faced with the prospect of drowning he found death was something he feared rather than embraced. For MacArthur, too, sailing round the world provided the stimuli, and the space, to rethink how modern consumer culture is configured, which ultimately led her to re-channel her energies away from sailing into something that she realized was more important and far-reaching: advocacy for a circular economy.[5] As she explains on her website:

> Sailing around the world against the clock in 2004, I had with me the absolute minimum of resources in order to be as light, hence as fast, as possible. At sea, what you have is all you have, stopping en route to restock is not an option and careful resource management can be a matter of life or death – running out of energy to power the autopilot means you can be upside down in seconds. My boat was my world, I was constantly aware of its supplies limits and when I stepped back ashore, I began to see that our world was not any different. I had become acutely aware of the true meaning of the word 'finite', and when I applied it to resources in the global economy, I realised there were some big challenges ahead.[6]

Spirituality transformed

Clearly, solitude at sea can offer opportunities for self-knowledge and reflection that rarely appear in the relentless activities in which we indulge on land. Perhaps it takes an experience like a life-threatening storm for us to discover whether, like Newton, we might find well-concealed religious faith in a life that seemed outwardly to protest the contrary, or whether, like Wesley, we will find limitations to a faith that, in his case, was so central a feature of his life. However, to seek to identify one's true self, and to imagine one's response as if, as Wesley put it, at 'every moment on the brink of eternity', is a discipline that may facilitate a focus on what is really important, wholly aside from the immediate pre-occupations of everyday life.

This, of course, leads to a further salient aspect of spirituality that may be brought into focus in the light of passing through the sea, namely the possibility of change and new life through an encounter with the prospect of death, or with our finitude, or merely with our insignificance before the vastness of the ocean.

Thus we see that the sea can be a place of transformation – where through facing death, life is brought new meaning and direction – and that this is not just a reality in individuals' lives in the modern world. Shakespeare already picked up on this aspect of the sea in some of his late plays, such as *Pericles*, *The Winter's Tale* and, of course, *The Tempest*. Here the sea, and in particular disastrous sea voyages, cause personal separation and apparent loss, but thereby also effect a change, or even a curative force, in the lives of certain key characters.[7] As it is famously expressed in *The Tempest*:

> Nothing of him that doth fade,
> But doth suffer a sea-change
> Into something rich and strange.

However, the idea of the sea as a place of transformation, and even of effecting a passage from death to life, has much deeper roots, since in the Old Testament it is in passing through the sea that Israel was formed as the people of God. It became the pivotal moment of faith and salvation that was the reference point for all that followed. From the New Testament, it is clear that passing through water (though not actually the sea) in baptism was not only the marking point for the beginning of Jesus' own ministry, but even in the earliest Church it became the key event for each convert as they passed from death to new life in Christ and received the Holy Spirit.

Spiritually transformed?

Two stories, one in the Old Testament and one in the New, show
the potential for spiritual transformation that encounters at sea
might effect, but also of limitations in the human capacity fully
to trust in God. The familiar tale of Jonah is the first and Peter
walking on the water is the second.

Jonah's story is one of fleeing God in disobedience but encoun-
tering God initially through a storm, and through miraculous
intervention, being thrown into the sea, then swallowed and
spewed out on to dry land by a 'big fish'. Jonah's spiritual transformation seems to be indicated by his willingness to be sacrificed
to save the others on the ship (Jonah 1.12) and by his prayer to
God for deliverance, once he is thrown overboard (Jonah 2.1–10).
Through the drama of sea storm and deliverance, a disobedient
prophet is transformed spiritually into an obedient prophet, ready
to respond to God's commands. He now understands what it
means to 'worship the LORD, the God of heaven, who made the
sea and the dry land' (1.9). An ironic feature of this story is that
once he has delivered his message of doom to the Ninevites and
they repent, Jonah is outraged that God should withhold the predicted judgement. He then needs to be taught a further lesson by
God in order to understand the value of divine compassion. It
would seem that even dramatic encounters might not lead to total
and lasting spiritual transformation, but rather often comprise
part of an on-going process through which spiritual change and
growth are achieved more gradually.[8]

Matthew 14.22–32[9] describes the experience of Jesus' disciples
being sent on ahead by him in a boat to cross the Sea of Galilee,
while he went up a mountain to pray. Their boat is buffeted by
waves as the wind is against them and then, in the middle of
the night, they see Jesus walking across the water. Uniquely in
Matthew's version of the events, Peter asks Jesus, 'Lord, if it is
you, command me to come to you on the water', to which Jesus
responds with the simple command, 'Come.' Peter, here acting
with faith, steps out of the boat and begins to walk on the water.
This is an extraordinary step of faith, yet in a very human fashion,

[30]... when he noticed the strong wind, he became frightened, and beginning to sink, he cried out, 'Lord, save me!' [31]Jesus immediately reached out his hand and caught him, saying to him, 'You of little faith, why did you doubt?' (Matthew 14.30–31)

This encounter shows Peter spiritually transformed initially, from being fearful to having the impulsive faith to step off the boat. Unfortunately, as Wesley found too, in circumstances of serious immediate danger, fear can rapidly triumph over faith. Both Wesley and Peter exemplify the paradox that they maintained their reliance on God even while their panicked behaviour showed that their dominant instinct was to be afraid and so discover a limit to their faith. Wesley found himself unable to face death; Peter lost confidence once he became aware of the storm around him and began to sink, yet his instinct was still to cry out, 'Lord, save me!'

This clinging to faith even amid fear and doubt is encapsulated by the words cried out by the father hoping against hope that Jesus might cure his epileptic (or demon-possessed) child: 'I believe; help my unbelief' (Mark 9.24). Paradoxically, this distressed and desperate parent realizes that his capacity to believe is limited, yet by imploring Jesus to help his unbelief simultaneously also recognizes that the limitation is his, not Jesus'. He both believes and is afraid to believe. This shows that there is a tension between faith and fear in many circumstances and experiences. Nevertheless, such occasions, as for Peter here and for John Wesley described earlier, can be steps on the way to a longer-term spiritual transformation where our hearts can become 'strangely warmed'[10] and a longer-lasting change is effected. As we read the Gospels and the Acts of the Apostles, we can observe that longer-term change in Peter's life.

Key message

Encounters with the sea can challenge and transform us in a multitude of ways – by its raw beauty and power; by its gift of enabling us to recognize our own insignificance; and by placing those who venture on to its waves into a situation of such immediate danger that it impresses upon the consciousness what is most salient in life in a way that is rarely possible in other circumstances. However, it also offers a window into the transcendent, impressing upon our senses a Power greater than ourselves that infuses all creation; it uplifts us by its beauty, bestowing us with a sense of connectedness through its immensity, and it forces upon us a sense of dependence on God as we surrender all illusions of control.

Challenge

Experience of the sea can provide a clearer window on to reality than is usually available to us. The challenge, then, is to live in the light of such experiences the sea, when the sea is not actually before us, imploding on our senses, overwhelming us with its beauty or threatening our very being. It was in a storm at sea that Newton discovered the grace of God, and Wesley recognized his fear of death; and it was through her solo circumnavigation of the globe that MacArthur had the insight to realize that her boat was like the world in microcosm in its finitude of resources. However, each of them discovered something that was no less valid in more secure and familiar circumstances, but just not otherwise visible to them. It took the challenge of the sea for these insights to impinge on their consciousness and to confront them with realities that may not otherwise have fallen within their horizon. Our challenge, then, is to enter imaginatively into that perspective, to know our frailty and the enormity of the world and to recognize our dependence on God while living under the illusion of autonomy and normality. It was a storm at sea that forced Wesley to consider 'what manner of men those ought to be who are every moment on the brink of eternity', but to think thus is much more difficult in the everyday. Paradoxically, tragic events such as terrorist attacks provoke an increase in people's social focus and sense of connectedness precisely because the prospect of imminent death impresses the supreme importance of relationships over the other concerns that occupy much of our lives.[11] Living with the same focus in the ordinary is much harder.

Reflection and discussion

- One theme that emerges repeatedly from the experiences of the writers discussed in this chapter is the life-changing impact of the prospect of death at sea. However, though not always contemplating this in a focused way, we all live with the limits of mortality and the preciousness of the one life on earth that we have been given. Preparing for the next life is much neglected in our time, but worthy of serious attention. Along with Wesley (notwithstanding the gendered language of his time), think about what 'manner of man' (or woman) you would wish to be and how you would wish to be found by your Maker. The idea of death as the lens through which a life might be evaluated also speaks to a secular perspective: how would you wish to look back (or for others to look back) on your life when your time comes? It is important not just to prepare for the next life but to live this one well, and indeed the pursuit of virtue arguably answers both imperatives. If you knew your time was very limited, what would become most important to you, how would your priorities change, and what would be the main focus of your life? Perhaps you should act on this now?
- The mariners portrayed in Psalm 107 recognized their utter dependence on God in their predicament at sea. Psalm 104, soon after celebrating marine life, reflects (in vv. 27–30):

> ²⁷These all look to you
>   to give them their food in due season;
> ²⁸when you give to them, they gather it up;
>   when you open your hand, they are filled with
>   good things.
> ²⁹When you hide your face, they are dismayed;
>   when you take away their breath, they die
>   and return to their dust.
> ³⁰When you send forth your spirit, they are created;
>   and you renew the face of the ground.
>
> (Psalm 104.27–30)

In other words, life and death, and all that sustains us, come from God. Despite our many technological developments, life and the many things that support it – like the earth's climate system – are beyond our control. Probably we pay insufficient attention to the contingent and dependent aspects of our existence. Think (or discuss) for a while what this means. Reliance on God may raise many issues, from divine grace and our responses to it, theodicy (the problem of God's justice in the face of suffering), gratitude and social justice (what do we own and should we have or share if everything we possess is a gift, ultimately not of our own making?). What does the giftedness of life suggest about our place in the world and our relationship to the rest of creation? Does the notion that we should tread lightly and view what we have as held in trust for the next generation suggest itself?

• Many of the spiritual impulses arising from contact with the sea explored here result from being attuned to nature, and absorption of its wild beauty and sensory impact. The loss of the richness of life through modern, often ultimately commercial, pre-occupations was recognized by W. H. Davies in his poem 'Leisure', published in 1911:

> What is this life if, full of care,
> We have no time to stand and stare.
> No time to stand beneath the boughs
> And stare as long as sheep or cows.
> No time to see, when woods we pass,
> Where squirrels hide their nuts in grass.
> No time to see, in broad daylight,
> Streams full of stars, like skies at night.
> No time to turn at Beauty's glance,
> And watch here feet, how they can dance.
> No time to wait till her mouth can
> Enrich that smile her eyes began.
> A poor life this if, full of care,
> We have no time to stand and stare.

William Wordsworth, in a classic example of the Romantic impulse in poetry, expressed this loss more vehemently in 'The World Is Too Much With Us':

The world is too much with us; late and soon,
Getting and spending, we lay waste our powers; –
Little we see in Nature that is ours;
We have given our hearts away, a sordid boon!
This Sea that bares her bosom to the moon;
The winds that will be howling at all hours,
And are up-gathered now like sleeping flowers;
For this, for everything, we are out of tune;
It moves us not. Great God! I'd rather be
A Pagan suckled in a creed outworn;
So might I, standing on this pleasant lea,
Have glimpses that would make me less forlorn;
Have sight of Proteus rising from the sea;
Or hear old Triton blow his wreathèd horn.

How far do you agree with Wordsworth's sentiment? Is his assessment of people's immunity to nature's charms accurate, and how might you address this in your own life? Some people see the Enlightenment and industrialization as ultimately stemming from Christianity itself, with its valuing of the individual and sense of purposefulness in history, giving rise to the commitment to 'progress'. Do you think this is a fair assessment or, indeed, even if it is true does it arise as a distortion of Christianity or one of its gifts? Does paganism really allow a better communion with nature as the poem suggests?

Action

Windows on to 'deep Eternity'[12] are not exclusively available at sea. William Blake, in his 'Auguries of Innocence', famously recognized that it is possible to gain insights into ultimate meaning through the smaller parts of the natural world:

To see a World in a Grain of Sand
And a Heaven in a Wild Flower,
Hold Infinity in the palm of your hand
And Eternity in an hour.

Urbanization and the pace of modern life have done much to
reduce our sense of connectedness with the wider web of life
and our openness to what it may reveal. However, we too have
the opportunity to experience the wonder of creation and access
a more profound level of contemplation. This can be achieved
through deep attentiveness to the natural world around us
(specifically, in the context of this book, the sea) and through the
wealth of images afforded us from memory, through the imagi-
nation or simply from those widely available online. St Francis of
Assisi is famous for his attunement to the natural world, achieved
through contemplative prayer in nature. This resulted in a deep
understanding that he himself was part of creation, and other
creatures simply his 'brothers' or 'sisters', but it also enabled
him to see the abundant goodness and presence of God in the
world. Take the opportunity, then, to sit quietly, perhaps for 30
minutes or so, and absorb the natural world around you, open to
the wonders of creation and the presence of God in it. Perhaps,
following Blake, you can contemplate something small such as
a wild flower or even a grain of sand. Maybe you could ensure
this way of praying becomes a regular part of your spiritual life,
among other disciplines.[13]

Notes

1 This and subsequent quotations are from the Journal of John Wesley.

2 The quotations here are from John Newton, 1765, *An Authentic
Narrative of Some Remarkable and Interesting Particulars in the Life of
Mr John Newton: Communicated in a Series of Letters to the Reverend
Mr Haweis, Rector of Aldwinckle, Northamptonshire, and by him (at the
Request of Friends) Now Made Public*, 3rd edition, London: printed for
S. Drapier, T. Hitch and P. Hill, Letter 8.

3 A solo long-distance yachtswoman, who in 2005 broke the world
record for the fastest solo circumnavigation of the globe.

4 The following quotations are taken from one of her autobiographical books, E. MacArthur, 2010, *Full Circle: My Life and Journey*, London: Michael Joseph (Penguin).

5 For the concept of a circular economy, see www.ellenmacarthur foundation.org/circular-economy.

6 See www.ellenmacarthur.com and www.ellenmacarthurfoundation. org – the latter website describes her work towards achieving a restorative and regenerative 'circular economy'.

7 The theme of the sea in literature is explored by W. H. Auden, 1967, 'The sea', in *The Enchafèd Flood or The Romantic Iconography of the Sea: Three Critical Essays on the Romantic Spirit*, New York: Vintage Books (Random House); see https://archive.org/details/enchafdfloodorroooaude. See also Bernhard Klein, 2002, *Fictions of the Sea: Critical Perspectives on the Ocean in British Literature and Culture*, Farnham: Ashgate Publishing.

8 The ministry of the prophet Jonah, son of Amittai, is also mentioned in 2 Kings 14.25, but it is unclear how this relates to the story in the book that is named after him.

9 With partial parallels in Mark 6.45–52 and John 6.16–21, but no parallel in Luke.

10 To quote John Wesley's description of his own ultimate spiritual transformation (Journal of John Wesley, Wednesday, 24 May 1738).

11 This is shown in the linguistic analyses of the social psychologist James Pennebaker and others; see, for example, Michael A. Cohn, Matthias R. Mehl and James W. Pennebaker, 2003, 'Linguistic markers of psychological change surrounding September 11, 2001', *Psychological Science*, 15/10, pp. 687–93.

12 This extract is taken from a poem by Emily Dickinson:

Exultation is the going
Of an inland soul to sea –
Past the houses, past the headlands,
Into deep Eternity!

Bred as we, among the mountains,
Can the sailor understand
The divine intoxication
Of the first league out from land?

13 Resources for 'Praying Nature with St Francis of Assisi', including some of his prayers, are available at www.praying-nature.com. For suggestions for praying the 'Examen' through nature, see www.patheos.com/Resources/Additional-Resources/Praying-with-Nature.

3

The God of the Sea and
All that Fills It

Created

Perhaps a good place to begin, when considering the God of the sea and all that fills it, is at the beginning – that is, with Genesis 1 and the story of creation. The key verses, about the sea and the creatures in it, relate to the third and fifth days of creation respectively:

> ⁹And God said, 'Let the waters under the sky be gathered together into one place, and let the dry land appear.' And it was so. ¹⁰God called the dry land Earth, and the waters that were gathered together he called Seas. And God saw that it was good
> ...
> ²⁰And God said, 'Let the waters bring forth swarms of living creatures, and let birds fly above the earth across the dome of the sky.' ²¹So God created the great sea monsters and every living creature that moves, of every kind, with which the waters swarm, and every winged bird of every kind. And God saw that it was good. ²²God blessed them, saying, 'Be fruitful and multiply and fill the waters in the seas, and let birds multiply on the earth.' (Genesis 1.9–10, 20–22)

These verses emphasize at least four things about God and his creation. First, it is clear that God considers the sea and the creatures in it as good. Since they are created prior to the creation of humans (Genesis 1.26–28), they have intrinsic value: they are important of themselves and not just in relation to humanity

– a theme that is explored in more detail later in the chapter. Second, the waters teem with life. There is an abundance and an extravagance to God's creation – God is profligate in creation. Third, the creatures of the sea receive the same command, 'be fruitful and multiply', that humanity too receives later (Genesis 1.22, 28). The sea creatures are to fill the sea just as humans are to fill the earth. Again God intends abundance. Finally, God creates every living thing with which the waters teem – from the smallest microscopic plankton to the largest of the sea creatures, the blue whale. The depth and breadth of his creation of the sea and the creatures in it arouse awe and wonder.

This awe and wonder has inspired artists and been captured in poetry, such as in the poem 'Fireflies of the Sea' by James Fenton:

Dip your hand in the water.
Watch the current shine.
See the blaze trail from your fingers,
Trail from your fingers,
Trail from mine.
There are fireflies on the island
And they cluster in one tree
And in the coral shallows
There are fireflies of the sea.

Look at the stars reflected
Now the sea is calm
And the phosphorus exploding,
Flashing like a starburst
When you stretch your arm.
When you reach down in the water
It's like reaching up to a tree,
To a tree clustered with fireflies,
Fireflies of the sea.

Dip your hand in the water.
Watch the current shine.
See the blaze trail from your fingers,
Trail from your fingers,

Trail from mine
As you reach down in the water,
As you turn away from me,
As you gaze down at the coral
And the fireflies of the sea.

This poem evokes the phenomenon of phosphorescence in the sea
caused by microscopic plankton, which emit light when they are
disturbed, as when the poet dipped his hand in the water. With
this introduction, let us move on to consider some other aspects of
God's sea and sea creatures that inspire awe and wonder.

Created to frolic

One of the most awesome experiences that one can have is to be
in a small research vessel in the middle of a storm in the North
Atlantic Ocean.[1] There both the beauty of God's creation and the
insignificance of human beings in the midst of that creation are
brought home with tremendous impact. Seeing waves taller than
the ship breaking around and on to the vessel makes it immediately apparent that we humans have little control over the forces
of creation. Being days from land, and off the main shipping
routes, underlines the feelings of isolation and of helplessness
should anything go wrong. It is therefore easy to identify with
Jesus' disciples' response when caught in a small boat in a storm
on the Sea of Galilee (Mark 4.35–41 and parallels).

At the same time, the awesome nature of God's creation is
brought home through experiencing the storm. There is the spectacular sight of the powerful waves breaking, and the awareness
of tremendous forces at work in the world, here exhibited in the
wind and the waves. When this experience is enhanced by the
presence of a group of pilot whales – both large adults and small
juveniles – swimming around the ship, the beauty of creation
is evident. The whales appear to be 'body surfing' the breaking
waves and this immediately brings to mind the reference in Psalm
104.26 to Leviathan, the great sea creature that God created to
'frolic' in the sea. The whales are incredibly well adapted to their

environment and very comfortable in it, in contrast to us humans bouncing around on the waves in our small 'tin can'.

Most of the time such behaviour, such 'frolicking', goes unobserved by human beings. Television programmes such as the BBC's *Blue Planet* and *Blue Planet II* have given us some insight into the life of ocean creatures, but have only captured a small part of what is actually happening in the open ocean. Only God truly sees such frolicking behaviour, which occurs out of sight of human eyes.

Denizens of the deep

A change of scene – now we are in the deep ocean, many kilometres below the surface on the abyssal plain, where no sunlight penetrates. Yet here, far below the ocean surface, life flourishes. A whale dies and falls to the seabed, and within a short time deep-sea creatures are feasting on this unexpected food from above. Such a whale fall may provide food for scavengers for up to two years. Once most of the soft tissue is gone, other deep-sea creatures can feast on the remaining organic matter and tissue left behind by the scavengers. Finally, bacteria come and attack the bones and extract food from them, in turn becoming food for sea snails and other creatures – a process that can take 50 to 100 years.

Moving away from the abyssal plain, we encounter the deep-sea ridges and trenches. Here water that has penetrated below the seabed and been heated geothermally emerges back into the ocean through hydrothermal vents (so-called 'black smokers'). The mineral-rich heated waters encounter the near-freezing waters of the ocean depths and chemicals are precipitated: having been in solution, they solidify into small particles, giving the appearance of a black cloud. These vents support a vast array of deep-sea creatures.

Normally, life on earth requires energy from sunlight. Plants use the power of the sun to produce sugars from carbon dioxide and water via photosynthesis. This energy is then passed on through the food chain to plant-eating animals and, in turn, to carnivores. In the deep ocean, however, the energy for life is supplied from the minerals and heat emerging from hydrothermal vents. This happens via a process called chemosynthesis, through which energy generated by chemical reactions is used to produce organic matter. Bacteria grow in this way, in due course becoming food for small creatures such as copepods (a group of small crustaceans), which in turn are eaten by snails, shrimps and fish. Many new and unusual species have been found in the vicinity of these deep-sea vents. Some biologists believe such vents provided the environment that allowed the first life to develop and flourish on earth – the so-called hydrothermal theory of the origin of life.

The deep ocean is a strange place, well outside the normal range of human experience and only recently explored by people using deep-sea submersibles. At great depths the pressure due to the overlying waters is tremendous, increasing roughly by one atmosphere for every 10 metres' increase in depth. Given that the world's oceans are approximately 4 kilometres deep on average, it is amazing to think that life can exist under pressures that would instantly squash a human being. Many of the creatures found at these depths look strange – and perhaps even ugly – to human eyes, but nevertheless evoke a sense of awe that they can exist – and even flourish – in an environment that is so hostile to human beings.

Ocean colour scene

Another change of scene – back to the ocean surface and the sunlit waters of a coral reef, teeming with life and colour. Coral reefs are places of abundant life, and are the most diverse of all ocean ecosystems. Anyone who goes snorkelling here is rapidly struck by the great profusion and variety of life, and by the beautiful colours of many of the creatures living there. Even more abundant than the fish are phytoplankton[2] (microscopic marine plants), which grow in the sunlit near-surface waters of the oceans. Like plants on land, phytoplankton use photosynthesis to grow, and as they contain chlorophyll (a green pigment essential for the absorption of energy from sunlight) they too are coloured green, just like grass or leaves on a tree. There are potentially millions of phytoplankton in a cubic metre of water, and this abundance is the basis of the oceanic food chain. They are the food for zoo-plankton[3] (microscopic and small animals), which in turn provide food for fish and ultimately even for human beings.

As phytoplankton contain chlorophyll, their presence changes the appearance of the ocean water from blue to green. This change can be detected using optical instruments (so-called ocean colour sensors) flown on satellites in space. These ocean colour scenes, images taken from hundreds of kilometres above the earth's ocean surface, reveal intricate patterns in the chlorophyll: whorls, swirls and streamers, generated by the currents and eddies in the ocean that move the phytoplankton around. On scales of 1 to 1,000 kilometres, these images reveal a beautiful and intricate pattern of planktonic life in the ocean, unseen by human eye until the advent of satellite observations from space.

Beauty and the beast

Perhaps the most startling thing about life in the ocean is how it can be both beautiful and yet also forbidding. Beasts such as the great white shark and the killer whale are dangerous and deadly predators. They are nonetheless striking and even beautiful in terms of their sleek body shape, which enables them to move through the water quickly when pursuing prey. How can creatures that are such ruthless killers be, at the same time, such a beautiful part of God's creation?[4] Watching them and similar sea creatures simultaneously evokes awe and horror, reminding us that God's creation is far from tame and safe.

Another creature that is perhaps not the most beautiful in appearance is the sea turtle. These animals seem somewhat ungainly as they haul themselves up a beach, but are far more graceful when swimming in the sea. In evolutionary terms, turtles pre-date humanity by tens of millions of years, and they can be found throughout the world's oceans with the exception of the polar seas. An idea of the scale of their endurance on earth is indicated by the geological changes they have survived. During their long period of existence on earth they have continued to use the same reproductive strategy, generation after generation, of returning year-on-year in large numbers to a beach to lay eggs. However, although they are famous for always returning to the same beach, when the modern sea turtle evolved about 110 million years ago, the continents were placed very differently from where they are now. Even the earth has changed beyond recognition during this time, and yet the turtles have still endured. Their long existence on the planet is now threatened by a variety of human activities such as fishing and beach developments. Sea turtles and other sea creatures are a good reminder to us that we humans are latecomers on the planetary scene.

The God of all creation

It is easy to slip into reading the Bible as the story of God's relationship with humanity (or indeed, more narrowly, with his chosen people) and his purposes for the human race. Of course, most of its pages do indeed centre around the experience and activity of quite a small group of people – in the Old Testament, the people of Israel and its sister nation Judah; and in the New Testament, Jesus and the early apostles. However, this is only one part of a much wider picture.

The broader context of the biblical narrative makes clear that there was a universal purpose in the particularity of choosing one people. Genesis 3—11 recounts a catalogue of human sin, culminating in the mixing of languages and the dispersal of the peoples across the earth after the building of the Tower of Babel. However, there immediately follows, from Genesis 12, God's plan to remedy this. He chose one man, Abram (later renamed as Abraham), through whom all peoples would be blessed. This promise was of course eventually understood to begin to find its fulfilment in the activity of the early Church, with the inclusion of the Gentiles. The affirmation that God created all of humanity and that his purposes are for all people provides an important context and reference point for the very particular, and usually quite localized, action detailed in the pages of the Bible. The 'camera' of the biblical authors may have focused on one particular people and geographical area, but it is the wider context that gives meaning to the picture in its totality.

However, there is an even more important aspect to the biblical narrative of God's dealings with his creatures, one that is provided by the frame around the whole story. The Bible begins not with humanity, but with the creation of heaven and earth and all that is in them (Genesis 1), and it ends again not with a focus on humanity, but with new heavens and a new earth (Revelation 21). This provides a vital context for the understanding of God's engagement with his creation. Although our experience of God is mediated through a human lens, and the Bible came into being in a particular corner of the earth, God's purposes are much wider than we often realize. Into this frame there has been inserted a story of God's engagement with his people, but the wider theological context invites us to imagine alternative pictures within that frame: the story of his relationship with sparrows or squirrels, or fish or whales.

Of course, the idea that God created everything and continues to be concerned for all his creatures is not confined to Genesis or to Revelation.[5] The Bible contains innumerable allusions to his interaction with creation, whether this is through the regulation of the seasons, the alternation of day and night, the provision of rain, or through ensuring the availability of food for particular species.

The response of creation: obedience and praise

The relationship between God and his creatures is not one-sided, for we also have many references to creation – both inanimate bodies such as mountains, seas or forests, and also animals – responding in praise and in obedience to him.

It is easy to take for granted the biblical notion that creation does God's bidding, whether (to take just a few examples) it is the earth putting forth vegetation at his command at the time of creation (Genesis 1.11), the stilling of a storm (Matthew 8.26; Mark 4.39; Luke 8.24; cf. Psalm 107.25, 29; Jonah 1.4–16), the bears coming out of the woods to maul the boys who mocked the bald Elisha (2 Kings 2.23–24), or the sea-serpent who will bite any evildoers who may attempt to escape the justice of God by fleeing to the bottom of the sea (Amos 9.3). Such incidents are often visualized from the perspective of God's purposes for humanity, of course, but this is not to deny how reciprocal the underlying relationship is. He tends to them, just as they respond to him.

The two-way interplay between God and creation is found in many parts of the Bible. God's wise provision for all his creatures is seen, for example, in Psalm 104, which we shall discuss further below. But this is balanced by the praise of God which bursts forth from creation back to the Creator, again most often in the psalms. A typical call to praise comes at the end of Psalm 69, beautifully encapsulating how universal the response to God is: 'Let heaven and earth praise him, the seas and everything that moves in them' (Psalm 69.34). This impulse to praise also finds fuller expression in Psalm 148 and in Psalm 96. Here is Psalm 148:

> ¹Praise the LORD!
> Praise the LORD from the heavens;
>   praise him in the heights!
> ²Praise him, all his angels;
>   praise him, all his host!
>
> ³Praise him, sun and moon;
>   praise him, all you shining stars!
> ⁴Praise him, you highest heavens,
>   and you waters above the heavens!
>
> ⁵Let them praise the name of the LORD,
>   for he commanded and they were created.
> ⁶He established them forever and ever;
>   he fixed their bounds, which cannot be passed.
>
> ⁷Praise the LORD from the earth,
>   you sea monsters and all deeps,
> ⁸fire and hail, snow and frost,
>   stormy wind fulfilling his command!

⁹Mountains and all hills,
　　fruit trees and all cedars!
¹⁰Wild animals and all cattle,
　　creeping things and flying birds!

¹¹ Kings of the earth and all peoples,
　　princes and all rulers of the earth!
¹²Young men and women alike,
　　old and young together!

¹³Let them praise the name of the LORD,
　　for his name alone is exalted;
　　his glory is above earth and heaven.
¹⁴He has raised up a horn for his people,
　　praise for all his faithful,
　　for the people of Israel who are close to him.
　　Praise the LORD! (Psalm 148.1–14)

As we can see, this joyful psalm systematically invokes all that
lives in each area of the three-part universe (heavens, earth and
seas) to 'praise the Lord' in a deliberately inclusive hymn of
praise. This begins with the heavens, including angelic and astral
beings, as well as the moon and the waters over the heavens (the
reservoirs for rain). It moves on to the great sea creatures and the
deeps and other meteorological phenomena that may have been
understood as stored over the horizon and therefore perhaps as
under the earth; and it ends with the familiar terrestrial world: the
mountains and hills, animals and birds, and varieties of human
life of all types.

One motivation for this praise is that God created his creatures,
established them permanently and ordered them in their proper
places, but another is his own majesty, which is 'above earth
and heaven' (Psalm 148.13). God is understood as intrinsically
worthy of creation's praise. A further cause for praise is God's
activity on earth, often in the human, or even merely national,
sphere. We see this in Psalm 148.14.[6] However, the human basis
for praise should not be over-emphasized, even if it appears to
be quite prominent. For example, in Psalm 148, the reference

to human concerns in v. 14 comes not as a ground for all creation to praise God (introduced by 'for' or 'because', as in 5b, 13b), but as further (and more personal) evidence of his greatness in addition to the testimony of creation.

Of course, the praise of God by all creation is not confined to the psalms: similar sentiments are found also in Revelation:

> [13]Then I heard every creature in heaven and on earth and under the earth and in the sea, and all that is in them, singing,
> 'To the one seated on the throne and to the Lamb
> be blessing and honour and glory and might
> for ever and ever!' (Revelation 5.13)

Such passages are a vital part of the biblical witness to the idea that not only did God create – and he continues to sustain – all creation, but creation itself witnesses to God's glory, and offers praise and obedience in response to him.

God of the blue planet

We have already seen that the Bible relates God's dealings with the world through human eyes (and words): although its narrative is set within the context of his creation and purposes for all that exists, much of its content is dependent on human traditions and experience centred around the life and faith of a small group of people in the Middle East. It contains many examples of untrammelled praise of the Creator by all creation, offered simply because he is a great and majestic God to whom it owes its being. Even so, the perspective of the biblical writers, who often wish to acknowledge their own gratitude to God for his involvement in their affairs, frequently colours its expression.

There is another interpretative lens, shared both by the writers and by modern readers of the Bible, and that is our 'terracentricity': we are creatures of the land and write from the perspective of the earth. Indeed, the ancient worldview passed down to us by biblical and classical writers knew of parts of Eurasia and Africa, but nothing of America, Oceania or the Antarctic. As a result, the world was envisaged as consisting of a single landmass, plus a few associated islands, surrounded by sea. The land was literally understood as at the centre of the earth, and the sea as more peripheral. Within the Old Testament, the Mediterranean is referred to as 'the Great Sea', while other seas, such as the Dead Sea, Red Sea and Sea of Galilee, were smaller; the Persian Gulf was remote and little known; and the Atlantic, though probably ventured into by Phoenicians trading with Cadiz, may as well have been in outer space. Thus, awareness of the ocean, as we would understand it, was limited. Even within the modern world, it is a commonplace that the surface of the moon has been better mapped than the sea floor, and the place of the sea in the Bible has still tended to be marginalized from our thinking.

However, if we re-examine the passages we have considered so far, it becomes obvious that the sea and its creatures are no less the object of divine concern than the rest of creation and that they participate fully in his praise and in obedience to him. Indeed, certain passages that celebrate the life of these creatures seem to go further and deliberately offer a challenge to our own self-centredness and self-importance, reminding us that God is God of all creation and we are but a small part in it.

The creation of the sea creatures

It is well known that in the account of creation in Genesis 1.1—2.4a, each day's work is concluded with the affirmation, 'And God saw that it was good', while on the sixth day, 'God saw everything that he had made, and indeed, it was very good.' Unquestionably, then, the Bible begins with a strong assertion both of the goodness of the individual works of creation and of the whole in its totality. However, if we look at the first act of creation of animal life, which is the creation of marine fauna, there are other points of interest that illuminate the place both of aquatic life and of animate beings more generally:

[20]And God said, 'Let the waters bring forth swarms of living creatures, and let birds fly above the earth across the dome of

the sky.' ²¹So God created the great sea monsters and every living creature that moves, of every kind, with which the waters swarm, and every winged bird of every kind. And God saw that it was good. ²²God blessed them, saying, 'Be fruitful and multiply and fill the waters in the seas, and let birds multiply on the earth.' ²³And there was evening and there was morning, the fifth day. (Genesis 1.20–23)

First of all, it is said that God '*created* [italics ours] the great sea monsters and every living creature that moves, of every kind, with which the waters swarm'. The verb translated 'created', *bārā*', is particular to the activity of God. A person may 'make' something, and God may too, as indeed he does the firmament (Genesis 1.7), the two great lights of the sun and moon (Genesis 1.16), the animals (Genesis 1.25) and human beings (Genesis 1.26). In fact, at the end of creation, he 'saw everything that he had *made* [italics ours], and indeed, it was very good' (Genesis 1.31). However, it is God alone who is said to 'create'. Genesis 1.1 opens boldly, 'In the beginning, when God created the heavens and the earth ...', and the same verb recurs at the end of this account, in Genesis 2.3, in which it is stated that 'God rested from all the work that he had done in creation'. Within this summarizing frame, *bārā*' is used of just two acts of creation. The first is the creation of human beings in Genesis 1.27. The verb *bārā*' occurs three times in this one verse, so is clearly emphatic:

> ²⁷So God *created* humankind in his image,
>   in the image of God he *created* them;
>   male and female he *created* them.
>
> (Genesis 1.27) [italics ours]

The other act of creation described in this way is prior to this, here in Genesis 1.21:

> So God *created* (*bārā*') the great sea monsters and every living creature that moves, of every kind, with which the waters swarm, and every winged bird of every kind. And God saw that it was good. (Genesis 1.21) [italics ours]

44

There are two important observations that may be made about this. First, the fact that the created nature of humanity is asserted in triplicate indicates that this is significant and has important theological content. The 'greater lights' are said only to be 'made', and this likewise suggests the downplaying of the importance of these supposed deities. (The sun and moon would have been regarded as gods in most ancient Near Eastern cultures, and indeed the words in Hebrew for 'sun' and 'moon' would have been the names also for these gods, so this was an important point to make in that context.) Conversely, then, 'creation' may imply particular divine involvement.

The second important point to make relates to the structuring of the creation account. The acts of creation have long been understood as falling into two main phases, namely those of separation (between day and night, earth and sky, sea and land, on days 1 to 3), and those filling the spaces now created (the astral and planetary bodies, plants and animals, from days 3 to 6). The first animals to be 'created' are the marine and avian species listed in Genesis 1.21, and the last is humanity, in v. 27. This could therefore be understood as implying special significance for the dominant species in each realm: the aquatic life and birds in the sea and sky respectively (v. 21), and humanity on the earth (v. 27). Possibly more likely, however, is that the two uses of the verb *bārā'* may be part of a framing device, which implicitly affirms the 'createdness' (rather than merely the making) of the entire animal kingdom, including the land animals mentioned in the intervening verses, to whom the verb *bārā'* is not directly applied.

There are two further aspects of the creation of sea life that need to be noted. First, the Hebrew words translated here in vv. 20 and 21 as 'living creature(s)' ('Let the waters bring forth swarms of living creatures ... So God created ... every living creature ... with which the waters swarm') are exactly the same as those employed in the following creation account, in Genesis 2.7. This verse describes how, after being formed from the dust and having the breath of life breathed into its nostrils, the first human became a 'living being'.[7] The expression is used in Genesis 1.24, in respect of the creatures of the earth, and in 1.30 of any 'living being' (NRSV, 'everything that has the breath of life'). Even in

Genesis 2.19,[8] it is used to describe the animals brought to the human to be named. This confirms that to be a 'living being' (or 'living creature') is not a mark of human uniqueness.

Finally, the aquatic creatures are blessed with the words, 'Be fruitful and multiply and fill the waters in the seas' (Genesis 1.22). This corresponds, of course, very clearly to the blessing of humanity in v. 28: 'Be fruitful and multiply, and fill the earth'. Although the great sea creatures and lesser, 'swarming', forms of marine life are not told to subdue the seas nor granted dominion over other species, the fundamental commissioning to this point is the same. The implication may be that human beings may be the dominant terrestrial species, but filling the waters is left to its own kind. To put it another way, humanity may be licensed to fill about 29 per cent of the earth's surface, but 71 per cent is reserved for others to populate and make their home.

The great sea creatures

The inhabitants of the sea are described in the NRSV of Genesis 1.21 as the 'great sea monsters' (though 'great sea creatures' might be a better translation) and 'every living creature that moves, of every kind, with which the waters swarm'. Clearly the ancient Hebrews did not have the benefit of modern biological categories and classifications, nor did they distinguish sharply between different classes of animal in the way we might. In particular, the line between known and unknown, and literal and metaphorical beasts, cannot be sharply drawn. This is a phenomenon already familiar to us from medieval maps, which were decorated with dragons and sea monsters along with other features that we would recognize as belonging to the real physical world.

The 'great sea monsters' are referred to by the Hebrew word *tannîn*, which seems to be a rather nebulous term, at times synonymous with 'serpent',[9] and at others apparently designating any large sea creature (as seems to be the case here in Genesis 1.21). Sometimes it may describe the crocodile (or what we might regard as a mythically enhanced crocodile, as in Job 41), and at other times it appears to be a fearsome beast of a primarily mythological character. In this latter case, *tannîn* is rendered into Greek as *drakōn*, which of course gives us the English 'dragon'.

However, there are also two named beasts, Leviathan and Rahab, which are occasionally described not merely as 'a *tannîn*' but as '*the tannîn*' ('the dragon, the sea serpent, the great sea creature'). This is true of Leviathan in Isaiah 27.1 and of Rahab in Isaiah 51.9. In both of these contexts, they seem to have a mythological and even symbolic character, representing Israel's (and even God's) enemies:

> Awake, awake, put on strength,
>   O arm of the LORD!
> Awake, as in days of old,
>   the generations of long ago!
> Was it not you who cut Rahab in pieces,
>   who pierced the dragon? (Isaiah 51.9)

> On that day the LORD with his cruel and great and strong sword will punish Leviathan the fleeing serpent, Leviathan the twisting serpent, and he will kill the dragon that is in the sea. (Isaiah 27.1)

Rahab seems to be a hostile figure also in Psalm 89.10 (though it is not obvious there that he is a 'dragon'). In Psalm 74.13–14, Leviathan is mentioned in connection with 'dragons', and again he represents the enemies of God or Israel, apparently at the Exodus. At times, Rahab and Leviathan may even present a threat to the created order: in Job 3.8, rousing Leviathan seems to be a means of destroying the day of Job's birth, while in Job 26.12–13 the killing of Rahab is associated with the alleviation of a threat to the heavens. Both Leviathan in Isaiah 27.1 and Rahab in Job 9.13 also appear to be personal enemies of God.

The situation is nonetheless complicated by the fact that Rahab is sometimes simply synonymous with 'Egypt', rather as we might speak of 'Britain' and 'the UK' (or 'America' and 'the USA') interchangeably. This is the case in Psalm 87.4[10] and Isaiah 30.7.[11] Leviathan may also not always be a mythical creature, but in some contexts features as a normal beast of the sea rather than a dragon. This is the case, for example in Psalm 104.26, where Leviathan is a playful creature and does not seem to have the symbolic significance we find in other passages in which he is mentioned. In any context, however, Leviathan may reasonably be understood to be a fearsome and dangerous beast.

We shall be returning to Rahab and Leviathan in Chapter 6. For the present purpose, though, we need to attend to two passages in particular that express God's delight in the awesome creature that is Leviathan, namely Psalm 104 and Job 41. Both of these important texts seem self-consciously to avoid the prevalent attitudes of human-centredness (anthropocentrism) and utility that so often beset our thinking in relation to the natural world. The first of these, Psalm 104, is a beautiful and wondering celebration of God's wisdom in creation and in his provision for his creatures.

Psalm 104

The psalmist here expresses awe and delight in the way God has firmly established the earth and ordered the natural world. He has set the pattern of day and night, of seasons and diverse habitats for different creatures, and continues to provide for its varied life through bestowing on it the means to water and food. The last aspect of God's work to be considered in the psalm is his wise creation and continuing provision for the creatures of the sea:

> 24O LORD, how manifold are your works!
>   In wisdom you have made them all;
>   the earth is full of your creatures.
> 25Yonder is the sea, great and wide,
>   creeping things innumerable are there,
>   living things both small and great.
> 26There go the ships,
>   and Leviathan that you formed to sport in it.
>
> 27These all look to you
>   to give them their food in due season;
> 28when you give to them, they gather it up;
>   when you open your hand, they are filled with good things.

[29]When you hide your face, they are dismayed;
  when you take away their breath, they die
  and return to their dust.
[30]When you send forth your spirit, they are created;
  and you renew the face of the ground. (Psalm 104.24–30)

The subject of Psalm 104.27–30 is a collective group referred to as 'these' or 'them'. This relates back most immediately to Leviathan and the full diversity of marine life, but it is almost certainly intended to concern all animate creation, as described in the preceding stanzas. All are dependent on God for their food, life, breath and creation.

Verse 24 also is a generalized exclamation: how great (or many) are the things that God makes (or does)! He has made them all in wisdom, a concept in Hebrew that includes the idea of purpose, order and goodness. In the Genesis creation account, there is an emphasis on the blessing that enables the creatures to multiply. Here, similar delight is taken in how the earth (which can mean the whole earth, as contrasted with the heavens, and not just the dry land) is full of God's creatures. The word translated 'creatures' here in v. 24, *qinyānîm*, is an unusual one in the Old Testament. It usually means 'property',[12] but may also at times have the implication of paternity.[13] It is worth pausing to reflect on this statement, since the ecological ramifications for taking seriously the idea of animals as God's special possession,[14] and even as his children, are huge.

The psalm illuminates the idea of the diversity and wise creation of God's own creatures by reference to the sea, which teems with innumerable and varied life, 'living things both small and great' (a clear 'echo' of Genesis 1.21). As we saw above, there are traditions in the Bible that present Leviathan as the enemy of God's people or even of God himself. Seen in this light, it is extraordinary how much the psalm diverges from this perspective. Far from being threatening to the created order, Leviathan has deliberately been placed in the sea as part of that order, in accordance with God's wise plan. In fact, from God's perspective, Leviathan is not a terrifying beast, but an innocent playful creature, made to enjoy the great, wide sea. Various English terms are used to translate the

word given here in the NRSV of v. 26 as 'sport', but the important element is the idea of playfulness, celebration, merrymaking
and enjoyment – and this can even include dancing.[15] Wisdom in
Proverbs 8.30–31[16] is said to 'delight' in God's creation, but here
the frolicking Leviathan is doing the same thing. The storm-waves
that highlight human vulnerability in the vast alien environment
of the sea are a perfect playground for great sea creatures such
as Leviathan (and indeed the pilot whales mentioned earlier), at
home in the habitat they were created to enjoy.[17]

Job 41

Psalm 104, then, recognizes that God's ways are not our ways,
and that his purposes and relation to his creation are beyond
our understanding and, indeed, not centred specifically on our
own welfare. This insight is taken to a new level in the book of
Job. Job, the chief character in the book, is an exemplary man,
'blameless and upright, one who feared God and turned away
from evil', as we are told in the opening verse. Yet he is suddenly
struck by disaster, losing all his possessions, his children, and even
his health. Most of the book is taken up with a debate between
Job and his three 'friends' about what the cause of his suffering
might be. The friends assume that the fault must be with Job,
but he steadfastly insists on his innocence and questions the
justice of God instead. The solution to this question of theodicy
(divine justice) is provided at the end of the book, in Job 38—41,
when God answers Job from the whirlwind and challenges Job
to respond. These poetic divine speeches, which conjure up the
wonders of the natural world and the divine wisdom behind it,
eloquently communicate the majesty and power of God, forcefully showing how the mystery of his ways is far beyond Job's
understanding.

The first of the two divine speeches (Job 38—39) reveals how
God's cosmic power is seen in the first acts of establishing the
cosmos and in his rule over the outer reaches of all that exists
(leading out the stars, riding the clouds, and knowing the way
even to the springs of the sea and the gates of death). It is also

shown in his superlative command of light and dark and over the full range of meteorological phenomena. However, as well as exposing the magnitude of the scale of God's action, the author also celebrates the details of his personal care for – and delight in – wild creatures, and his provision for fearsome predators and beasts devoid of human purpose. In a similar way, in his wisdom he bestows rain on the desert in order to enable the grass to grow where there are no people.

These themes – the magnitude of God's activity in creation, but also his personal care for his creatures – are developed further in the second divine speech (Job 40—41). Here, God expresses his delight and pride in two of his most terrible creations, Behemoth ('the Beast') and Leviathan, both of which are unfavourable to human wellbeing and devoid of any practical use. Behemoth broadly resembles the hippopotamus, while Leviathan bears the closest similarities to the crocodile. Since our present interest, though, is on the sea, we shall focus on Leviathan, who is the more impressive of the two and also appears (according to Job 41.31-32) to inhabit a marine environment:

> [1]"Can you draw out Leviathan with a fish-hook,
>> or press down its tongue with a cord?
> [2]Can you put a rope in its nose,
>> or pierce its jaw with a hook?
> [3]Will it make many supplications to you?
>> Will it speak soft words to you?
> [4]Will it make a covenant with you
>> to be taken as your servant for ever?
> [5]Will you play with it as with a bird,
>> or will you put it on a leash for your girls?
> [6]Will traders bargain over it?
>> Will they divide it up among the merchants?
> [7]Can you fill its skin with harpoons,
>> or its head with fishing-spears?
> [8]Lay hands on it;
>> think of the battle; you will not do it again!
> [9]Any hope of capturing it will be disappointed;
>> were not even the gods overwhelmed at the sight of it?

¹⁰No one is so fierce as to dare to stir it up.
　Who can stand before it?
¹¹Who can confront it and be safe?
　– under the whole heaven, who?
¹²'I will not keep silence concerning its limbs,
　or its mighty strength, or its splendid frame.
¹³Who can strip off its outer garment?
　Who can penetrate its double coat of mail?
¹⁴Who can open the doors of its face?
　There is terror all around its teeth.
¹⁵Its back is made of shields in rows,
　shut up closely as with a seal.
¹⁶One is so near to another
　that no air can come between them.
¹⁷They are joined one to another;
　they clasp each other and cannot be separated.
¹⁸Its sneezes flash forth light,
　and its eyes are like the eyelids of the dawn.
¹⁹From its mouth go flaming torches;
　sparks of fire leap out.
²⁰Out of its nostrils comes smoke,
　as from a boiling pot and burning rushes.
²¹Its breath kindles coals,
　and a flame comes out of its mouth.
²²In its neck abides strength,
　and terror dances before it.
²³The folds of its flesh cling together;
　it is firmly cast and immovable.
²⁴Its heart is as hard as stone,
　as hard as the lower millstone.
²⁵When it raises itself up the gods are afraid;
　at the crashing they are beside themselves.
²⁶Though the sword reaches it, it does not avail,
　nor does the spear, the dart, or the javelin.
²⁷It counts iron as straw,
　and bronze as rotten wood.
²⁸The arrow cannot make it flee;
　slingstones, for it, are turned to chaff.

²⁹Clubs are counted as chaff;
   it laughs at the rattle of javelins.
³⁰Its underparts are like sharp potsherds;
   it spreads itself like a threshing sledge on the mire.
³¹It makes the deep boil like a pot;
   it makes the sea like a pot of ointment.
³²It leaves a shining wake behind it;
   one would think the deep to be white-haired.
³³On earth it has no equal,
   a creature without fear.
³⁴It surveys everything that is lofty;
   it is king over all that are proud.' (Job 41.1–34)

There are several aspects of the portrayal of Leviathan that are striking. The first is the enormous emphasis on the impossibility of human control over this beast: it can neither be tamed nor killed and traded (Job 41.1–11), and indeed so fierce is it that even the divine beings are afraid of it (Job 41.9, 25).[18] Nonetheless, God takes pride in this creature, revelling in its unrivalled strength and invincibility (vv. 12–34). In fact, the language of this passage seems exaggerated in its celebration of Leviathan's superlative qualities. Not only does Leviathan have extraordinary defensive attributes, such as its closely overlapping impenetrable scales and powerful mouth, well-bestowed with 'terror all around its teeth'; it also has more offensive characteristics: it breathes fire and smoke (vv. 18–21) and raises itself up with such a great crashing that even the gods are terrified (v. 25). As the intense rhetorical questions accumulate throughout this passage, the reader gains a strong impression of how God himself remains untouched by the limitations experienced by his other creatures before Leviathan. He alone, then, is able to enjoy this most awesome creature. Even the deep is affected by Leviathan's activity (vv. 31–32), yet its creator positively delights in its power and ferocity, enthusing, 'I will not keep silence concerning its limbs, or its mighty strength, or its splendid frame' (v. 12). An important aspect of this speech, then, is that Leviathan has an intrinsic value before God, against which its perception by human beings is irrelevant.

Another implicit message – which was spelled out already in

relation to the other great semi-mythical beast, Behemoth, at the very opening of this divine speech (Job 40.15) – is clearly understood to apply equally to Leviathan: 'I made [him] just as I made you'. This message is taken to a new level in the closing verses of the speech in Job 41.

> [33]On earth it has no equal,
>   a creature without fear.
> [34]It surveys everything that is lofty;
>   it is king over all that are proud. (Job 41.33–34)

The perspective of the second divine speech begins with implied equality between God's most fearsome beasts and human beings ('I made [him] just as I made you'). However, by the end it has shifted to suggest the apparent superiority of Leviathan (effectively: 'There's nothing like him') and to acknowledge his exalted status. Leviathan himself can look on everything or everyone who is lofty and exalted. This phrase might include those held in dignity and honour[19] who are of high status, or the proud, or simply anything that is high (like mountains and trees). He is even king over all the majestic (or proud) creatures, which are probably (in view of Job 28.8, where they are mentioned again)[20] particular wild animals.

To state that Leviathan is the object of Yahweh's[21] purposeful creation, and that God provides for him and takes special pride in him as one of his greatest creative achievements, is an extraordinary claim to make. From a human perspective, this creature is highly dangerous and threatening, an enemy from whom God's protection might be sought, yet nonetheless it has a deliberate place carved out for it in God's well-ordered world. The clear implication is that God's purposes are not wholly centred on humanity, nor are they accessible to the human mind.

Creator, creation and creatures

It is easy to forget that we human beings are not the be-all and end-all of God's magnificent creation. From one perspective, we are simply creatures in it. From another perspective, we are unique in his creation in being made in the image of God (Genesis 1.27).[22] However, as our consideration of both the beauty and abundance of ocean life and of relevant biblical passages has shown, the oceans and the life in them are of intrinsic value to the Creator. Too often we have seen ourselves as the pinnacle of God's creation and lords and masters (note the masculine nature of this terminology) of all we survey. The creatures of the sea remind us that God takes delight in his creation irrespective of the presence of humans in it.

For most of human history we have at best 'dabbled our toes' in the great waters of the ocean, and the majority of human activity at sea has taken place within sight of land.[23] We have remained largely unaware of the beauty and complexity of the life in the ocean, which has nevertheless been a delight to its Creator. God takes pride and pleasure in aspects of creation that we may not even be aware of – or, in the case of the ocean, may have been unaware of for most of human history. As Shakespeare so eloquently put it (through Hamlet's words to Horatio): 'There are more things in heaven and earth, Horatio, than are dreamt of in your philosophy.'[24] Our human understanding and knowledge are limited and it is good to remember that!

Stepping back from our own immediate concerns we see that God is profligate in his creating, as the abundance of life in the sea shows. He has concerns with his creation and plans for it that transcend what we can imagine (cf. Ephesians 3.20).[25] Even creatures that appear ugly or horrifying still have their place, along with other life that we can recognize as exquisitely beautiful. This should chasten us and remind us that our whole experience of, and perspective on, life may be too human-focused, too centred on terrestrial creatures, thus missing the grandeur of who God is and what he has created. We should be mindful that 'the earth is the Lord's and all that is in it' (Psalm 24.1), including the oceans and sea creatures, and it is not ours to do with as we wish. As Christopher Wright has put it so well, 'Trashing someone else's property is incompatible with any claim to love the other person.'[26] If we claim to love God then a degree of humility is called for in how we live on the planet that we share with the rest of his creatures, from microscopic life through to blue whales (the largest creatures on the earth). The challenge is to become aware of the richness of God's creation – to know that it is important to him and was not solely created for us human beings to exploit and misuse.

Key message

God is bigger and his concerns are much broader than human beings tend to imagine. The Bible, and indeed our whole experience and outlook, may be human- and land-centred, but God is not confined to this perspective. He values and nurtures all of creation, receiving praise and obedience from all his creatures. An examination of key passages concerned with marine creatures maintains their status, like humans, as 'living beings' created by God, and blessed with the gift of filling the oceans. Even the greatest sea creatures, which are so much an object of human fear, have a valued part in his creation, being formed to enjoy their freedom in the wide expanse of the sea. Indeed, within at least one strand of biblical thought, even a quasi-mythological beast that is thought of as highly inimical to human life and as causing terror among the divine beings, is the object of God's special pride and pleasure and is envisaged as having a prime place over his creatures. A humble re-centring of our limited, self-centred and anthropocentric perspective is called for.

Challenge

It is clear that humans are having an enormous impact on the sea and the creatures in it. Sometimes this is due to deliberate actions, as is the case with overfishing, and sometimes it is inadvertent, such as the problems that arise with discarded fishing nets and plastic shopping bags, which then entangle or are swallowed by creatures such as sea turtles, leading to their death.

Other impacts are a mixture of deliberate and accidental, as in the case of pollution. Humans use the sea for waste disposal deliberately (for example, sewage) but also accidentally, such as when rain and rivers wash fertilizers used in agriculture into the sea. This addition of nutrients can cause harmful blooms of algae that produce poisonous toxins. These toxins can then enter the food chain, affecting marine life and even causing the death of higher species such as dolphins.

Ecosystems that have flourished over long periods of time are being rapidly disrupted by human activity. More worryingly, this disruption is happening at an ever-increasing pace. Some of these systems have developed over thousands and even millions of years prior to the advent of humanity on the planet. Now they are having to change in response to human impacts on timescales of years to decades, much faster than they may be able to adapt successfully in order to survive.

Reflection and discussion

- God's concern for sea creatures is only one facet of his love for all of creation. How should we live if we are to emulate and respect that love? What changes in our lives might that involve?
- We tend to imagine God's concerns as being fairly contiguous with our own. Imagine for a minute Job's God, who is beyond human understanding and operates on a cosmic stage within which we comprise an insignificant part. How should we respond to this God, and does this biblical perspective make us revise any of our own goals and priorities?
- Another aspect of God seen in the Bible is that he is present and active throughout creation, watching over the world and providing for all his creatures. Consider the following poem, *The Maldive Shark* by Herman Melville:

About the Shark, phlegmatic one,
Pale sot of the Maldive sea,
The sleek little pilot-fish, azure and slim,
How alert in attendance be.
From his saw-pit of mouth, from his charnel of maw,
They have nothing of harm to dread,
But liquidly glide on his ghastly flank

Or before his Gorgon head;
Or lurk in the port of serrated teeth
In white triple tiers of glittering gates,
And there find a haven when peril's abroad,
An asylum in the jaws of the Fates!
They are friends; and friendly they guide him to prey,
Yet never partake of the treat –
Eyes and brains to the dotard lethargic and dull,
Pale ravener of horrible meat.

The poem describes the mutual relationship between sharks and pilot fish. The latter eat parasites on the former and bits of food left by them. Small pilot fish have been observed swimming into sharks' mouths to clean away fragments of food from between their teeth. In turn, the sharks provide protection for the pilot fish from predators. Does this poem and the discussion of Psalm 104 above (particularly vv. 27–29, which refer to God giving food to all the creatures 'in due season', so that they are 'filled with good things') qualify or challenge the common perception of 'nature' as being 'red in tooth and claw'? If the feeding of dangerous carnivorous predators is part of the wise provision of God (and indeed essential for balanced and healthy populations of wild species), how should this inform our actions and reactions?

Action

A huge challenge inhibiting us from recognizing and addressing the environmental damage to the ocean is that to a large extent we are isolated from it as we have little contact with the sea on a regular basis. We may go to the beach on holiday but we are largely unaware of the life teeming in the sea and our impact on it. We may become more aware of marine creatures through television programmes, most notably the BBC's *Blue Planet* and the various other series in which sea life has featured (for example, *Planet Earth* and *The Living Planet*), but that rarely leads us to consider our responsibilities towards them. However, this is not an excuse for complacency. Therefore, the question arises: what response can I (or we) make? What action should I (or we) take?

Unfortunately, there are no easy or straightforward answers. There are, however, some simple things that we can all do. Possibly the best way to start is to allow the biblical perspective on sea life to inform our thinking. The passages we have looked at affirm God's delight in the diversity of creatures in the sea, great and small, and his intention that they should be fruitful and multiply and fill the waters. This might move us to pray not just for human situations, but for the creatures of the ocean too, and to take into fuller account the welfare of marine life when making decisions about our own behaviour.

Perhaps our greatest opportunity to make a difference, at least in a direct sense, is as consumers. What are we buying and where does it come from? How was that beautiful shell caught? Which fish in our supermarket come from a sustainable source? There are a number of websites that document and certify which species of sea creature have been caught in a way that is not damaging to the ocean, and which have populations that are not overfished.[27]

An important aspect of this is not just the welfare of fished species, so that this happens sustainably, but about damage to marine habitats (as through bottom trawling). A further problem is that of bycatch – that is, creatures accidentally caught in nets but which are not saleable. This can involve fish that are of the 'wrong' species (even if dead and in principle edible), but sadly it can also entail the accidental netting of mammals such as dolphins or seals, or of birdlife, such as albatrosses, several species of which are now becoming endangered as a result of entrapment in fishing equipment. Attention to labelling can help us establish whether food is not just sustainably fished, but instead is 'dolphin friendly' or caught by one of the less damaging methods (such as line fishing).

Perhaps a little more challenging would be to try to buy food that has been grown without the use of large amounts of fertilizer, such as organic food, as this is perhaps less easily identified. Linking to this is a need to reduce our use of plastic shopping bags (easily done, as there are many alternatives). These simple things may not solve all the problems, but they are ways in which we can all contribute to their solution. This may come at a cost to us personally in terms of finance, as these choices are likely to increase the amount we spend on food shopping, for example. As in many other things, there is a sacrificial element in trying to live in the world and care for it as God would want us to do.

Notes

1 This section is based on the experiences of one of us (Meric) in carrying out oceanographic research in the North Atlantic.

2 Phytoplankton from the Greek *phyton* = light (cf. photon) and *planktos* = drifter (because plankton are carried by the ocean currents).

3 Zooplankton from the Greek *zōon* = animal (cf. zoo) and *planktos* = drifter.

4 As an aside, we note that the same comment can be applied to human beings – both ruthless killers (at times) and (at others) a beautiful part of God's creation.

5 Among the many references to God's creation of all that is, including heaven, earth, sea and all that live in or on them, see Exodus 20.11, Nehemiah 9.6, Psalm 146.6, Acts 4.24, 14.15, Revelation 10.6, the briefer expression in Jonah 1.9, and the fuller account in Proverbs 8.22–31. The references to 'all things' in John 1.3, Colossians 1.16 and Hebrews 1.2–3 may also be compared. The ongoing aspect of God's continuing activity in creation is expressed especially powerfully in Psalm 104.

6 A further example comes in Psalm 69. Here the call for all creation to praise God is followed, in a way that might surprise us, with the words:

³⁵For God will save Zion
  and rebuild the cities of Judah;
  and his servants shall live there and possess it;
³⁶the children of his servants shall inherit it,
  and those who love his name shall live in it. (Psalm 69.35–36)

Psalm 96 is even more intriguing. We shall focus on the last three verses, since these are the ones that relate to our theme:

¹¹Let the heavens be glad, and let the earth rejoice;
  let the sea roar, and all that fills it;
¹²let the field exult, and everything in it.
  Then shall all the trees of the forest sing for joy
¹³before the LORD; for he is coming,
  for he is coming to judge the earth.
  He will judge the world with righteousness,
  and the peoples with his truth. (Psalm 96.11–13)

Almost the same words are quoted also in Psalm 98.7, Isaiah 42.10 and 1 Chronicles 16.32.

7 '[T]hen the LORD God formed man from the dust of the ground, and breathed into his nostrils the breath of life; and the man became a living being.' (Genesis 2.7)

8 'So out of the ground the LORD God formed every animal of the field and every bird of the air, and brought them to the man to see what he would call them; and whatever the man called every living creature, that was its name.' (Genesis 2.19)

9 For example, in Isaiah 27.1.

10 Among those who know me I mention Rahab and Babylon;
Philistia too, and Tyre, with Ethiopia –
'This one was born there,' they say. (Psalm 87.4)

11 For Egypt's help is worthless and empty,
therefore I have called her,
'Rahab who sits still.' (Isaiah 30.7)

12 This may be seen from the remaining occurrences, which are found in Genesis 31.18, 34.23, 36.6, Leviticus 22.11, Joshua 14.4, Psalm 105.21, Proverbs 4.7 and Ezekiel 38.12–13.

13 This is true of the verbal form in Genesis 4.1, where there is a word play between the name 'Cain' (*Qayin*, in Hebrew) and Eve having 'begotten' him (*qānîthî*). Note how in English, too, there is a linguistic connection between 'getting' and 'begetting'. For the meaning 'create', see Genesis 14.19, 22, and probably also Proverbs 8.22; also compare Psalm 139.13.

14 The verbal form of this word is used in some contexts to describe the crucial moment at the Exodus when God (be)got, acquired, purchased or created Israel. See Exodus 15.16, where it refers primarily to their 'purchase' (or acquisition) out of slavery, by which they became his possession, a process more familiarly described as 'redemption'. Indeed, the two words, translated in the NRSV as 'acquired' and 'redeemed', are used together in Psalm 74.2 to encapsulate the election and deliverance of the people from Egypt. (The close connection between 'purchase/acquisition' and redemption is spelt out in Jeremiah 32.7, which speaks of the 'right of redemption by purchase', and is illuminated further in Ruth 4.4–5, where both terms are again employed.) The same verb is also often used elsewhere to describe the purchase of slaves: see, for example, Leviticus 22.11, 25.44–45, 50, Deuteronomy 26.68 and Ecclesiastes 2.7. The person who has thus made a purchase is not just a buyer, but becomes the 'owner' (see Isaiah 1.3, referring to an ox's owner), so this further illuminates the relationship encapsulated in God's 'getting' of Israel. The people of Israel become God's. Nonetheless, when referring to the bringing of Israel out of slavery/exile, the exchange of money is clearly not in direct view: it is the taking possession or reclaiming of the people that is the important aspect, as may be seen especially in Isaiah 11.11. At the same time, the same verb also appears to refer specifically to

Israel's creation in Deuteronomy 32.6, as its association with the motifs of being 'made' and 'established' makes clear. This background needs to be borne in mind when we try to understand the 'creatures' in Psalm 104.24 also as God's special possession and as made by him.

15 It is also the same verbal root that gave Isaac his name, from Sarah's laughter.

16 ³⁰then I [Wisdom] was beside him, like a master worker;
and I was daily his delight,
rejoicing before him always,
³¹rejoicing in his inhabited world
and delighting in the human race. (Proverbs 8.30–31)

17 The Bible contains another reference to creaturely playfulness, in Job 40.20, in which the mountains are described as 'where the wild animals [literally, 'living things of the field'] play'.

18 The NRSV reads 'gods' in both of these verses, but some other translations do not. The Hebrew of v. 9 is actually rather difficult and unclear, so although it may refer to 'a god' it could simply concern an unspecified person. As a result, many English versions use an impersonal form of the verb or speak of 'one' or 'a man', not a god (or gods) at all; for example, the NIV has 'the mere sight of it is overpowering', while the NKJ reads, 'Shall one not be overwhelmed at the sight of him?'. Verse 25 has much clearer Hebrew: it is just the interpretation of it that is disputed, which is why some translations say that it is 'the mighty' who are afraid. The debate hinges on the word *ēlîm*, which usually means 'gods', as in Exodus 15.11, Daniel 11.36, and as in the expression 'sons of the gods' in Psalms 29.1 and 89.6 (v. 7 in the Hebrew). The word (in the plural as well as the singular) can also be applied to the God of the Old Testament. The reason why some translations refer to 'the mighty' partly stems from discomfort with the idea of the Bible referring to 'gods'. But it is also because there are a few instances where the word *ēlîm* may be used adjectivally to describe a god-like quality, that of being 'mighty' or 'great'. Psalm 36.6 could refer to the 'mighty mountains' or to the 'mountains of God', and Psalm 80.10 to 'mighty cedars' (or 'cedars of God'). The problem is that this meaning is disputed, and *ēlîm* does not elsewhere stand alone to refer to 'mighty ones' instead of qualifying another noun (as in 'mighty mountains') or simply meaning 'God' or 'gods'.

19 An example would be the Servant of the Lord in Isaiah 53.13, who 'shall be very high'.

20 This passage in Job 28 refers to human mining, something that birds and wild animals are not part of:

⁷That path no bird of prey knows,
and the falcon's eye has not seen it.
⁸The proud wild animals have not trodden it;
the lion has not passed over it.

21 'Yahweh' is the personal name for God found in the Bible and is usually translated in English versions (out of deference to the Jewish belief that it should never be uttered) as 'the LORD'. In older Bibles and hymns it is sometimes given as 'Jehovah', but this form has generally fallen out of use. The earliest Hebrew Bibles consisted only of a consonantal text – they included just the consonants, and no vowels (though some consonants did double duty to indicate which might be intended). Later, vowels were added, but the scribes placed the vowels for the word 'the Lord' (actually, 'my Lord', *Adonai*) around the letters Y-H-W-H to indicate that 'Adonai' should be said instead of God's name being spoken. The result looked like Yeh[o]wah, from which the English Jehovah was mistakenly derived. 'Yahweh' is the accepted reconstruction of the divine name, but many Bibles retain the convention of reading 'the LORD'.

22 The meaning of being made 'in the image of God' is much debated. Most think that it does not simply concern the external appearance of human beings, but something more intrinsic in their nature, for example, spiritual capacities, especially the ability to relate to God. Another influential explanation is that it means that human beings are God's representatives on earth, perhaps acting to some extent on his behalf. In this case, it could be closely related to the idea of 'dominion' and would qualify it: whatever dominion is, it must express the idea of acting in a God-like way in relation to the rest of creation, entailing responsible care and rule.

23 Even most naval battles have been fought within sight of land. Think of the Spanish Armada or Trafalgar (the latter named after a nearby landmark Cape Trafalgar near Cadiz). An exception was the Midway carrier battle in the Pacific Ocean during World War Two.

24 Of course, when Shakespeare wrote, the term 'philosophy' would have included natural philosophy, what we would call 'science' today.

25 Here reference is made to 'him who by the power at work within us is able to accomplish abundantly far more than all we can ask or imagine'.

26 C. J. H. Wright, 2006, *The Mission of God: Unlocking the Bible's Grand Narrative*, Nottingham: Inter-Varsity Press, p. 414.

27 A good website to start with is www.fishonline.org. The Marine Conservation Society also has a 'Good Fish Guide': see www.mcsuk.org.

4

Human Creatures and the Life of the Sea

Gone fishing ...

The Gospels relate that Jesus' first disciples were fishermen from Galilee (Mark 1.16-20). They were familiar with working in a sometimes dangerous environment (Mark 4.35-41), but such a way of life was necessary to bring food to the table. Fish were part of the diet of the Galileans as shown by the account of the feeding of the 5,000 (John 6.1-13), where the boy contributed five loaves and two fish that were the basis of the miracle. Jesus clearly ate fish, as we can see from the story of his appearance to the disciples after his resurrection (Luke 24.36-43) and he also prepared fish for his disciples to eat in another resurrection appearance (John 21.1-14). In fact, after the shock of Jesus' death on the cross and his resurrection, the disciples had gone back to the security of what they knew best – fishing (John 21.1-3). These few examples from the New Testament reflect the familiar interactions that humans have had with life in the sea over thousands of years – fishing, then cooking and eating the fish that have been caught.

Some of this interaction is captured in the traditional eighteenth-century sea shanty 'Song of the Fishes':

Come all ye bold fishermen, listen to me,
And I'll sing you a song of the fish of the sea.

Chorus
Blow ye winds westerly, westerly blow;
We're bound for the southward, so steady we go.

First comes the bluefish a-wagging his tail;
He comes upon deck and yells: 'All hands make sail!'

Next come the eels, with their wagging tails,
They jumped up aloft and loosened the sails.

Next jump the herrings, right out of their pails,
To man sheets and halyards and set all the sails.

Next comes the porpoise, with his stubby snout,
He jumps on the bridge and yells: 'Ready, about!'

Next comes the swordfish, the knight of the sea,
The order he gives is 'Helm's a-lee!'

Then comes the turbot, as red as a beet,
He shouts from the bridge: 'Stick out that foresheet!'

Having completed these wonderful feats,
The blackfish sings out next to: 'Rise tacks and sheets!'

Next comes the whale, the largest of all,
Singing out from the bridge: 'Haul mainsail, haul!'

Then comes the mackerel, with his striped back,
He flops on the bridge and calls: 'Board the main tack!'

Next comes the sprat, the smallest of all,
He sings out: 'Haul well taut, let go and haul!'

Then comes the catfish, with his chuckle head,
Out in the main chains for a heave of lead.

Next comes the flounder, quite fresh from the ground,
Crying: 'Damn your eyes, chucklehead, mind where you
    sound!'

Along comes the dolphin, flapping his tail,
He yells to the boatswain to reef the foresail.

Finally the shark, with his three rows of teeth,
He flops on the foreyard and takes a snug reef.

Then up jumps the fisherman, stalwart and grim
And with his big net, he scoops them all in.

This illustrates the diverse range of sea creatures that live in the sea and that humans have caught over the centuries. However, its comical, even somewhat callous, tone raises the question as to whether this is an appropriate way of thinking about the sea life to which it refers: it is disconcerting to imagine these creatures willingly working to man the sailing ship, eagerly serving the fishermen's purposes only to be netted by their human masters in the end. There is no hint of compassion, nor of respect for these creatures. In this chapter we explore what the Bible has to say about how humanity interacts with the creatures of the sea and how we might respond to that understanding.

Us and them

Much of the life in the ocean seems strange and exotic to humans. Watching television programmes like *The Blue Planet* emphasizes even more the difference between our lives and that of the creatures in the ocean. Oceanic life is also distant from us physically and so does not really impinge on our day-to-day awareness or existence, beyond visiting the fish counter at the local supermarket. This leads to a classic 'us and them' situation where we do not pay much regard to those, whether they are human or non-human creatures, who are different from us. They are 'other' and strange and therefore beyond our concern, or even awareness. This is seen especially in the food choices of pesco-vegetarians, who avoid eating land-based creatures but will still consume fish.

Unfortunately, though we might think oceanic creatures are beyond our concern, for the first time in human history we are having a significant impact on oceanic life. This occurs not just indirectly, through rising ocean temperatures and acidification,[1] but more directly. Here we illustrate three such direct impacts:

commercial fishing, oil spill disasters, and the drying out of the Aral Sea. These show the major and possibly irreversible impacts of humanity on the ocean, that part of the earth which makes it look like a beautiful blue marble hanging in space.[2]

Hoovering the ocean

Commercial industrial-scale fishing vessels can 'hoover up' many tons of fish from the ocean, significantly depleting fish stocks in large areas to the extent that they may never recover. Overfishing happens when the number of fish caught exceeds those being added to the stock through reproduction. It is thought that as many as 85 per cent of the earth's fisheries may be fully- or over-exploited or depleted. Despite the imposition of fishing quotas, fish stocks can take many years to recover. In some areas the combination of overfishing and naturally occurring phenomena can lead to a disastrous collapse of fish stocks. Such a collapse occurred in the Peruvian anchovy fishery due to the combination of the 1972–3 El Niño[3] and overfishing, with the stocks not recovering until the 1990s. When the 1997–8 El Niño occurred, the Peruvian government imposed fishing restrictions so, while the stocks dropped, they also recovered more quickly. In contrast, the collapse of the Grand Banks cod fish stock, leading to a fishing ban by the Canadian government in the 1990s, was attributable to human over-exploitation.

Attempts to manage over-exploitation and move to more sustainable fishing practices have met with mixed success. Even if legitimate fishing vessels keep to the rules (for example, quotas, or minimum fish size caught), there are less scrupulous people who run illegal fishing vessels and are happy to ignore the rules for a quick profit. There is evidence that illegal fishing off the coast of Somalia, leading to loss of fishing income to local coastal communities, led to the rise of piracy in that area of the ocean. The consequences of illegal fishing impact both the fish and people (often poorer people) who rely on them for food and for their livelihood.

Apart from the depletion of fish stocks, with long-term

consequences for feeding humanity, the removal of large quantities of fish from the ocean changes the balance of the oceanic ecosystems in ways that are not fully understood. For example, eliminating a top predator, such as tuna, will have consequences all down the food chain. Similarly, eliminating a prey species, like anchovy, will affect the predatory fish further up the food chain. In either case the consequences of such disruptions on ecosystems that have taken millions of years to evolve are difficult to predict. We are changing these systems on timescales of years to decades, much faster than they can evolve and adapt in response.

Oil on troubled waters

The phrase 'pouring oil on troubled waters' is meant to suggest a calming influence. Unfortunately, in recent history, humans have been pouring crude oil into the ocean, so causing trouble rather than preventing it. Our desire for fossil fuel oil, which is necessary to many industries (for example, for transport and the manufacture of some plastics), has led to greater sub-seabed oil extraction and the transport of ever larger quantities of crude oil around the world in so-called super-tankers. Inevitably things go wrong – disasters happen and crude oil ends up in the ocean.

Perhaps the most dramatic example to date of direct ocean oil pollution caused by human beings is the Deepwater Horizon drilling platform disaster in the Gulf of Mexico in 2010 – the largest marine oil spill ever. This is just the latest of a series of major oil spills that span the last 50 years, going back to such incidents as the *Torrey Canyon* oil tanker spill in 1967. Such disasters directly affect the life in the oceans and in coastal areas where the oil often ends up on beaches. The Deepwater Horizon event[4] released almost five million barrels of crude oil into the Gulf of Mexico, about 20 times more than the *Exxon Valdez* oil spill off Alaska in 1989. It is considered the worst environmental disaster in US history. The ecological consequences of the Deepwater Horizon event were severe, affecting many species, including several already endangered ones: dolphins, several types of turtle, fish, molluscs, crustaceans and deep-water corals. Certain effects, such as deaths of some creatures, were immediate, but there appear to have been longer-term impacts too. The oil not only affected ocean life but also marine birds, as well as coastal areas and beaches and the flora found there, such as mangroves and marsh grasses. Years after the event, scientists are finding continuing impacts on oceanic flora and fauna and the area around the seabed oil well is lacking in much of the marine life that was there prior to the spill.

Therefore, somewhat ironically, not only is the burning of fossil fuels affecting the ocean negatively,[5] but the extraction and transport of the fuels themselves are also leading to negative consequences. On a smaller scale, ships that clean out their tanks at sea add to the pollution problem. The adage 'oil and water do not mix' might be better re-worded as 'oil and water *should not* mix', at least in the ocean.

Left high and dry

The graphic image of a camel walking past a ship stranded on the sand of what was once part of the bottom of the Aral Sea underlines our human ability to cause enormous and detrimental changes to the earth. For the first two-thirds of the twentieth century, the Aral Sea was the earth's fourth largest inland sea.[6] In the 1960s the Soviet government started diverting the two main rivers that fed the sea to irrigate the surrounding desert. By 2007 the Aral Sea had shrunk to 10 per cent of its original size (which was 26.300 square miles) and split into smaller lakes. The eastern part of the sea has now completely dried up and is known as Aralkum Desert! The drying up of the Aral Sea has been described as one of the planet's worst environmental disasters.

The result of all this is that the once prosperous fishing industry has disappeared, with a major impact on the locals' livelihood. The diversion of the rivers that fed the Aral Sea was done to enable the growing of cotton in the surrounding areas, and this worked for a while. Unfortunately, many of the irrigation canals were poorly built and much of the water in them was lost through leakage or evaporation, so the growing of cotton became problematic. The region is also heavily polluted due to the run-off of pesticides and fertilizers from the land, leading to health problems. The little that remains of the sea has become increasingly salty, and indeed it now has about three times the salinity of the seawater in the oceans,[7] which has devastated many of the ecosystems that might otherwise have survived in a reduced form. Where the sea has dried up entirely, this has left a residue of salt, fertilizer and other pollutants on what was once the seabed. The dried-out seabed is also the source of increasing dust storms in the region, which deposit salt on to the surrounding land, affecting the ability of people living there to grow crops, and impacting on their health too.

More recently there have been initiatives to restore the Aral Sea. There has been some success in the North Aral Sea due to the efforts of the Kazakhstan government and funding from the World Bank. In contrast, the South Aral Sea has largely been abandoned and allowed to become a desert, and is now being explored for oil. It would appear that humanity's 'god-like' powers have principally caused large-scale disasters rather than enhancing the beauty and majesty of the natural world. Sadly, the pattern of human use leading to the drying of major lakes is perpetuated throughout the world, with the Dead Sea in Israel and Jordan, Lake Chad in Africa, Lake Poopo in Bolivia and many others also being among the casualties.[8]

A right to lord it over other creatures?

All discussions seeking a Christian or biblical perspective on the relation of human beings to other animals at some point refer back to Genesis 1.26, 28:

[26]Then God said, 'Let us make humankind in our image, according to our likeness; and let them have dominion over the fish of the sea, and over the birds of the air, and over the cattle, and over all the wild animals of the earth, and over every creeping thing that creeps upon the earth ...' [28]God blessed them, and God said to them, 'Be fruitful and multiply, and fill the earth and

subdue it; and have dominion over the fish of the sea and over the birds of the air and over every living thing that moves upon the earth.' (Genesis 1.26, 28)

Much of what we have described – the drying of inland seas as a result of the diversion of rivers, the accidental killing of fish through oil spillages, and overfishing – may be attributed to human domination of the natural world, whether over the earth, broadly conceived, or more directly over fish themselves. But can this really be what the Bible intends? Perhaps not surprisingly, there has been considerable debate over what these verses mean. Do they provide the answer to our current ecological crisis, or may they rather be blamed for the exploitative attitude of many people, especially in the West, towards the rest of creation over the many generations since the Enlightenment. The literature on this is vast, and it would be impossible to evaluate all of what has been said on this topic here, so instead we shall confine ourselves to three main points.

In the image of God

There has been much discussion about what it means for humanity to be made in God's image. Does this refer to our physical appearance, to our spiritual capacity, or to how we should act? However, one thing is clear and that is that to be in any way like God in our relation to the rest of creation is to exercise a caring, protective, responsible role. We have fallen a long way short of this and any efforts we may make will be flawed. Nonetheless, the general nature of what it should be like to emulate God is not in dispute, however much the detail of the best course of action in a specific situation may be subject to debate.

On another level, this description could also be taken as a warning. We may not feel comfortable with this, but the reality is that human beings, especially after the dawn of the industrial age, have exercised an impact on the earth that is unprecedented and global in its effects. So much is this the case that scientists now speak of the 'Anthopocene' age, comparable to previous geological eras, but defined by the massive and utterly unprecedented impact of humans on the climate and environment. This sets us apart from other species in terms of what we may achieve, often for ill rather than good, but it also has serious implications with regard to the responsibilities that attend our actions. If our behaviour is god-like in its scale and impact, it is clearly incumbent on us to be God-like in how we make decisions and exercise our powers.

The wider perspective

It is important to be aware that Genesis 1.26, 28, though much quoted, only offer one aspect of biblical thinking about our relation to the rest of creation. We have already in the previous chapter considered the perspective offered by the book of Job, which suggests the high value placed by God on non-human creatures and even on those that are detrimental to us, and this calls into question any presumption that humanity must take centre-stage in God's purposes. At its best, the Bible is able to express wonder at creation for its own sake, freed from utilitarian illusions and even going so far as to celebrate those creatures that may potentially be detrimental to human safety (Psalm 104). This sense of creatureliness in creation and coexistence rather than mastery is one that provides a model which, though often outside the modern horizon, is increasingly recognized as necessary for ecological flourishing.

More immediately, Genesis 1 needs to be read together with the second creation account in Genesis 2—3. Whereas Genesis 1.26–28 offers a vision of humanity as having a unique relationship with God, made in his image, Genesis 3 by contrast dwells on human fallibility and mortality, as creatures who come from the dust and return to dust. Of course, there is truth in both portrayals, which is exactly why they have been recorded side by side. However, this also means that, when we try to construct a theological anthropology – a theology of humanity – we need somehow to take both of these insights into account and to balance them appropriately. We are sinful and fallible and can struggle to make our way in the world, and yet there is also a promise of blessing and a unique relationship with God, as well as the responsibilities that attend that.

The other biblical passage inviting comparison with Genesis 1 is Psalm 8. It speaks of human beings as 'made ... a little lower than God, and crowned ... with glory and honour', and as well as referring again to dominion over other creatures (Psalm 8.5–8). However, what is most striking about this psalm is its tone of wondering celebration as it beautifully expresses a sense of unwonted privilege at the place of humanity in the created world:

> ³When I look at your heavens, the work of your fingers,
>   the moon and the stars that you have established;
> ⁴what are human beings that you are mindful of them,
>   mortals that you care for them? (Psalm 8.3–4)

This sense of being exalted beyond one's due invites the recognition of a real responsibility in relation to the rest of the living world, and an attitude of awe and humility rather than arrogance and pride.

Within the wider perspective of the Bible as a whole, it can be seen that the original creation of Genesis will be renewed and restored in the new creation (cf. Isaiah 65.17–25; Revelation 21.1—22.5).⁹ In this it is apparent that Jesus' death and resurrection is not just about the restoration of human beings to a right relationship with God, but about the renewal and restoration of the whole of God's creation (Romans 8.19–22; Colossians 1.15–20, in which we note the stress on *all things*). In light of this, we should live now in the anticipation of the coming new creation and with an awareness that God's concerns transcend merely human-centred ones.¹⁰ This awareness too leads to a degree of awe and humility.

Vulnerable humanity

However, perhaps the most important (and most neglected) perspective in which to read Genesis 1.26, 28 is in relation to the experience of those to whom it was addressed. The Bible is replete with references reflecting the idea that wild animals are something to be scared of, and expressing the hope that its people might be safe from them. This most probably underlies the mention in Mark 1.13 that when Jesus was in the wilderness tempted by Satan 'he was with the wild beasts; and the angels waited on him'. He was in this unsafe region, and yet preserved and ministered to.

The much-quoted vision of the peaceable kingdom in Isaiah 11.6–9 pairs the most feared predators with young domestic livestock and snakes with little children:

> 6The wolf shall live with the lamb,
>   the leopard shall lie down with the kid,
> the calf and the lion and the fatling together,
>   and a little child shall lead them.
> 7The cow and the bear shall graze,
>   their young shall lie down together;
>   and the lion shall eat straw like the ox.
> 8The nursing child shall play over the hole of the asp,
>   and the weaned child shall put its hand on the adder's den.
> 9They will not hurt or destroy
>   on all my holy mountain;
>   for the earth will be full of the knowledge of the LORD
>   as the waters cover the sea. (Isaiah 11.6–9)

This is a vision not just of 'peace' but of safety, of a world in which it might be possible to live without fear of being hurt or killed or having one's livestock ravaged by wild beasts. Fear of snakes (and indeed lions) is something that recurs at various points in the psalms. However, the most revealing expression of the hardships of human existence comes in Genesis 3.14–19:

> 14The LORD God said to the serpent, '...
> 15I will put enmity between you and the woman,
>   and between your offspring and hers;
> he will strike your head,
>   and you will strike his heel.'
> 16To the woman he said,
>   'I will greatly increase your pangs in childbearing;
>   in pain you shall bring forth children,

yet your desire shall be for your husband,
  and he shall rule over you.'
¹⁷And to the man he said,
 'Because you have listened to the voice of your wife,
   and have eaten of the tree
 about which I commanded you,
   "You shall not eat of it,"
 cursed is the ground because of you;
   in toil you shall eat of it all the days of your life;
¹⁸thorns and thistles it shall bring forth for you;
   and you shall eat the plants of the field.
¹⁹By the sweat of your face
   you shall eat bread
 until you return to the ground,
   for out of it you were taken;
 you are dust,
   and to dust you shall return.' (Genesis 3.14–19)

The three curses encapsulate the fears and struggles of the ancient people to whom we owe the Bible: danger from wild animals; the pain and difficulty of childbearing; and the struggle to eke out a living from the soil. These are the mirror image of the blessings of Genesis 1.26, 28, which (addressing these very issues in the same order) promise that people shall have dominion over all creatures; that they shall be fruitful and multiply; and that they shall subdue the earth. We have benefited from these blessings superlatively well, despite the genuineness of the struggles expressed in Genesis 3, which are still a reality for many today. However, the promise that we shall ultimately be safe and flourish as a species, notwithstanding the many challenges to our existence, is a far remove from the aspirations to domination and exploitation that have marred our interactions with other creatures to date.

Of course, if Genesis 1.26, 28 should be read as providing a blessing that enabled growth from the initial point of creation to the development of a flourishing human society that was distributed over the land and able to master its environment sufficiently to ensure survival and modest prosperity ('blessing'), then this limited aim creates few problems. This is especially so

if we recognize that it has long since reached a point of fulfilment, past which a stable state, rather than an ever-increasing
population and GDP, is envisaged. In this case, our present-day
over-consumption and exploitation amounts to inappropriate
greed.[11] Thus, if it is read within our own cultural framework as a
licence either to assume a selfish consumptive attitude to creation
or the apparent narcissism of kingly rule, this passage is deeply
troubling. However, if it is taken, more appropriately to its original social setting, as a promise of ultimate flourishing attained
through a process extending over many generations, despite the
challenges of existence, much of the sense of threat is dissipated.

Thankfully, the general picture provided by Genesis 1.26, 28
in respect of humanity's relationship to all other creatures can be
supplemented by many specific references to human interactions
with fish and sea life, most obviously when engaged in fishing.

Fishing in the Gospels

Of course, fishing is a major part of the backdrop to Jesus'
ministry, and the suggestion that the disciples will henceforth
'fish for people'[12] or incidents such as the miraculous catch of
fish (positioned in Luke's Gospel in connection with the call of
the fishermen disciples,[13] and in John as a defining part of Jesus'
final resurrection appearance)[14] indicate a value system that was
affirmative of this way of life. Jesus was clearly not a vegetarian,
and indeed fish would have been a staple food and the primary
source of protein in first-century Galilee. It is an interesting reflection of this that Jesus is described as giving thanks or blessing and
distributing fish[15] more than he is wine,[16] although the blessing
and breaking of bread is mentioned slightly more than either.[17]
Given the custom among the first Christians of sharing food, it is
perhaps not surprising then that from a very early time fish were
depicted in association with the Eucharist.[18] Of course, the fish
became one of the first Christian symbols, with each of the letters
of the Greek word *ichthus* (*i-ch-th-u-s*) providing the initial letters
of the cypher *Iēsous Christos Theou [H]uios Sōtēr* (Jesus Christ,
Son of God, Saviour).

Nonetheless, this is not to say that in the Bible fish were regarded simply as a commodity with no intrinsic value of themselves. For a start, fishing is not presented as easy and hence Jesus' assistance comes after a night of fruitless labour and failure (Luke 5.5; John 21.3). In John's Gospel, Jesus is portrayed as waiting on the beach for his disciples by a 'charcoal fire ..., with fish on it, and bread' (John 21.9), and as inviting them, 'Come and have breakfast' (John 21.12). The fish satiate the disciples' hunger after a night at sea, just as the boy[19] who offers his five loaves and two fish provides the basis for the feeding of a crowd of 5,000 hungry men as well as women and children,[20] and, on another occasion, seven loaves and a few fish suffice to feed 4,000.[21] In the case of the feeding of the 4,000, both accounts state that the crowds had been with Jesus three days with nothing to eat, and there is concern that if they are sent home, they may faint on the way.[22] Here, each fish counts and answers a real need for sustenance, despite the possibility that earnest attempts to catch them may fail. It presents a way of living that is the antithesis of a modern culture characterized by maximum-tonnage fishing and high-waste consumption. This is not to dismiss the benefits of modern technology, but to highlight how ready and excessive availability affects our relationship with food and the value placed on a single fish.

A miraculous gift

There are two accounts in the Gospels of Jesus enabling a miraculous catch of fish, in Luke 5.1–11 and John 21.1–14:

[3]He ... taught the crowds from the boat. [4]When he had finished speaking, he said to Simon, 'Put out into the deep water and let down your nets for a catch.' [5]Simon answered, 'Master, we have worked all night long but have caught nothing. Yet if you say so, I will let down the nets.' [6]When they had done this, they caught so many fish that their nets were beginning to break. [7]So they signalled to their partners in the other boat to come and help them. And they came and filled both boats, so that they began to sink. [8]But when Simon Peter saw it, he fell down at Jesus'

knees, saying, 'Go away from me, Lord, for I am a sinful man!' [9]For he and all who were with him were amazed at the catch of fish that they had taken; [10]and so also were James and John, sons of Zebedee, who were partners with Simon. Then Jesus said to Simon, 'Do not be afraid; from now on you will be catching people.' [11]When they had brought their boats to shore, they left everything and followed him. (Luke 5.3–11)

[1]... Jesus showed himself again to the disciples by the Sea of Tiberias; and he showed himself in this way. [2]Gathered there together were Simon Peter, Thomas called the Twin, Nathanael of Cana in Galilee, the sons of Zebedee, and two others of his disciples. [3]Simon Peter said to them, 'I am going fishing.' They said to him, 'We will go with you.' They went out and got into the boat, but that night they caught nothing.

[4]Just after daybreak, Jesus stood on the beach; but the disciples did not know that it was Jesus. [5]Jesus said to them, 'Children, you have no fish, have you?' They answered him, 'No.' [6]He said to them, 'Cast the net to the right side of the boat, and you will find some.' So they cast it, and now they were not able to haul it in because there were so many fish. [7]That disciple whom Jesus loved said to Peter, 'It is the Lord!' When Simon Peter heard that it was the Lord, he put on some clothes, for he was naked, and jumped into the sea. [8]But the other disciples came in the boat, dragging the net full of fish, for they were not far from the land, only about a hundred yards off.

[9]When they had gone ashore, they saw a charcoal fire there, with fish on it, and bread. [10]Jesus said to them, 'Bring some of the fish that you have just caught.' [11]So Simon Peter went aboard and hauled the net ashore, full of large fish, a hundred and fifty-three of them; and though there were so many, the net was not torn. [12]Jesus said to them, 'Come and have breakfast.' Now none of the disciples dared to ask him, 'Who are you?' because they knew it was the Lord. [13]Jesus came and took the bread and gave it to them, and did the same with the fish. [14]This was now the third time that Jesus appeared to the disciples after he was raised from the dead. (John 21.1–14)

The primary theological significance in each case lies in the suggestion of Jesus' prophetic (God-given) level of knowledge and perception. According to Luke, for Peter this provided a moment in which Jesus' identity is suddenly revealed: Peter's sense of sin before Jesus is that of a man kneeling before his God (Luke 5.8). Thus, the fishermen's reaction is one of astonishment and even fear (vv. 9–10). Likewise, in John, it is when the catch was so great that they were not able to haul it in that the beloved disciple has the sudden realization, 'It is the Lord!'. The word 'Lord', however, was not only a conventional term of respect appropriate for a disciple addressing his master (akin to the English 'Sir'), but it was also the Greek translation of the Hebrew divine name 'Yahweh', which appears in our Bibles by the capitalized 'the LORD'. This name now in this moment is applicable equally to Jesus (John 21.7). In John too, then, the miraculous catch was an event through which Jesus 'showed himself' (v. 1), and in Luke it elicits not only recognition (Luke 5.8), but an immediate response of commitment: once the boats were brought in, the fishermen left everything and followed him (v. 11).

Such a catch is also demonstrative of the generosity of God, and indeed Jesus in John goes on to offer hospitality on the beach with the fish he has already set cooking, supplemented by some from the catch (John 21.9–13). In Luke it is also apparently an act of reciprocity after the fishermen have given Jesus the use of their boat as a platform for his teaching. In John it is a compassionate response to human need and weakness after their own efforts have left them still having caught nothing after a whole night of effort (John 21.3–5), and the same predicament also provides the background to the catch in Luke (5.5). The aspect of human dietary need is amply communicated in John by the question: 'Children, you have no fish, have you?' (John 21.5) and by the provision of breakfast on the shore.

However, if fish are among the good things on offer from God, it is also important to be attentive to the scale of this astounding miracle: the exact number of fish caught is not mentioned in Luke, but in John they are 'large fish, a hundred and fifty-three of them; and though there were so many, the net was not torn' (John 21.11). In Luke, the catch may be envisaged as somewhat larger, since the nets began to break and the men needed help from their partners in a second boat to bring the net in (Luke 5.6–7). Nonetheless, if two boats were filled to sinking point, hauling the fish in was still something accomplished, according to the inference of the narrative, by four people. Clearly, there is a huge technological difference between the fishing methods that provided the context for what was then regarded as such an unprecedentedly large catch, and those of today; it must also be recognized that within the limits of what was possible, the nets were filled to maximum capacity. Despite this, the scale of modern commercial fishing, in which a large fleet of trawlers might process tonnes of fish in a single day, and in which global marine capture production amounted to about 80 million tonnes in 2012,[23] is called into question by such a huge disjunction in the scales of fishing applicable in each case. If a catch of 153 fish is a moment of surprise generosity from God, this perspective prompts us to ask whether the millions of tons caught in today's commercial fishing can be anything other than hubristic greed, a rapacious grasping at something that is not humanity's to take, and that puts at risk the order and balance of creation.

When enough is enough

The attitude to fish shown in the Gospels is complemented by a brief reference in Numbers 11.22. Immediately prior to this verse, the Israelites have been complaining to Moses in the wilderness about the lack of meat, and God angrily promises to give them plenty for a whole month 'until it comes out of your nostrils and becomes loathsome to you' (Numbers 11.20). Moses' response is incredulity: there are 600,000 people.[24] He asks rhetorically, 'Are there enough flocks and herds to be slaughtered for them? Are there enough fish in the sea to be caught for them?' (v. 22).[25] This, of course, implies a strong denial: of course there is not sufficient to feed that number of people.[26]

From the perspective of the meeting of ancient and modern cultures through this one verse, the words of Moses present a challenge to what we now know to be the fallacy of assuming that the sea has limitless capacity to give. With modern fishing techniques, the fact that there are not enough fish in the sea to feed everyone is something that has to be confronted if they are going to be able to continue feeding many at all.

When disaster befalls the fish

In addition to the appreciation that fish are a limited and God-given resource, there are also a few passages in the Old Testament that present the plight of people suffering from invading forces or other disasters under the metaphor of fishing. This is important because it shows compassion for the experience of the fish, even while fishing was an important means of subsistence. We shall look at two examples here.[27] The first is from Ecclesiastes 9.11–12:

> [11]Again I saw that under the sun the race is not to the swift, nor the battle to the strong, nor bread to the wise, nor riches to the intelligent, nor favour to the skilful; but time and chance happen to them all. [12]For no one can anticipate the time of disaster. Like fish taken in a cruel net, and like birds caught in a snare, so mortals are snared at a time of calamity, when it suddenly falls upon them. (Ecclesiastes 9.11–12)

Ecclesiastes 9.12 likens human beings being snared at the 'evil time' that will suddenly fall upon them to fish being caught in an evil net or birds being held in a snare. The point is, as the preceding sentence states, that 'no one can anticipate the time of disaster'. The verb applied to the fish, rendered here in the NRSV as 'taken', denotes the idea of being seized or caught, with the added sense of being taken possession of. They are caught and held in the net, then taken away by the hunter. Particularly striking here is the use of the adjective describing the net in which the fish are caught (the 'cruel net'). In most contexts elsewhere, this word is translated simply as 'bad' or 'evil', a choice that implies a strong value-judgement about fishing as seen from the perspective of the victim. Clearly, the analogy is used simply to highlight the horror of the time that is to come. But this nonetheless reveals a capacity for understanding the plight of the hunted animal. Without this, the metaphor could neither have been forged nor have rung true to the audience.

A fisherman's merciless greed

The vulnerability of the fish is presented in a slightly different form in Habakkuk 1.14–17. This passage constitutes a metaphorical portrayal of the fate of the mass of humanity before a great conqueror who is intent on destroying one nation after another, with the peoples being likened to fish, and the conqueror to an insatiably greedy fisherman:

> [14] You have made people like the fish of the sea,
>   like crawling things that have no ruler.
> [15] The enemy brings all of them up with a hook;
>   he drags them out with his net,
>   he gathers them in his seine;
>   so he rejoices and exults.
> [16] Therefore he sacrifices to his net
>   and makes offerings to his seine;
>   for by them his portion is lavish,
>   and his food is rich.
> [17] Is he then to keep on emptying his net,
>   and destroying nations without mercy?
>   (Habakkuk 1.14–17)

Here the fish are described as having no ruler, which seems to imply that they are lacking direction and protection. As a result, they can put up no defence against the enemy – the ruthless fisherman – with his rod and seine who appears in the next verse. His rejoicing and exulting at his success is clearly felt to be odious. We have here the question, 'Will he therefore keep emptying his net and continue to kill the nations without pity?' – and this clearly implies that there is a proper limit to what should be taken.[28] The underlying point is evidently about a political and military situation facing Habakkuk's audience. Nonetheless, the passage reveals a capacity for perspective-taking that perceives the ugliness of excessive greed as it may be manifested in the exploits of a rapacious fisherman – and it clearly also recognizes the innate vulnerability of aquatic life.

So, the biblical passages on fishing that we have just looked at acknowledge the finitude of fish as a food source and show compassion for them as vulnerable creatures in the face of human rapacity. There is therefore much in this with which we can concur, not least as having now reached a point of serious depletion of fish stocks through the greed that is here condemned.[29]

Humanity and fish: a biblical perspective

We have (if the pun can be excused) trawled through many passages in the search for a biblical perspective on how we should relate to sea life. Fortunately, however, the picture appears to be reasonably consistent. References to fishing, although showing the importance of fish for the diet and economy of first-century Palestine in particular, also reveal an appreciation of fish as a gift from God, a modest idea of plenitude, and a realization of what it might feel like to be a netted fish, even if fishing was necessary. Further important strands are a horror of the excessive, gratuitous taking of fish (albeit with a metaphorical significance in thinking of the insatiability of power-hungry empires seizing ever more territory) and an instinctive appreciation that fish stocks are not limitless. We might also mention that in the Gospels the reality of hunger and the importance of sharing and responding to others' needs are also identifiable strands in the stories relating to fish.

More broadly, the blessing of Genesis 1.26, 28 needs to be read in the light of the Old Testament sense of human finitude and vulnerability before creation. It needs to be blended with Genesis 2—3's realization of human fallibility, Psalm 8's awe-struck feeling of undeserved privilege, and the responsibility that attends any claim to being in the image of God. Many of the themes discussed here are vividly explored in Robert Lowell's disturbing poem, 'The Quaker Graveyard in Nantucket. For Warren Winslow,

Dead at Sea'. Lowell's poem was dedicated to his cousin, lost at sea in World War Two, but much of the poem explores themes not just of human loss and vulnerability but, more profoundly, of the corruption, greed and arrogance that are understood as the underlying cause of deaths at sea. A very striking feature of the poem is that it begins with the quotation of Genesis 1.26, which functions as an epigraph, so that the composition as a whole should be read in the light of this verse and its encapsulation of humanity's responsibility and proper relationship towards the rest of creation. Particularly important for our present discussion is its treatment of whaling and the religious imagery and theological understanding attending this. The poem is too long to quote in full (though it is very much worth exploring in its entirety), so only the most pertinent sections are reproduced here. The bulk of the poem is set in a Quaker graveyard in Nantucket (hence the title), which provides the context for a meditation on death at sea, centred largely on the whale trade that was the focus of these sailors' activities. The poem is full of biblical and literary references, which are enriching but also require explanation, so some brief notes drawing attention to these are given at the end of the chapter.

> Let man have dominion over the fishes of the sea and the fowls of the air and the beasts of the whole earth, and every creeping creature that moveth upon the earth. (Genesis 1.26)

> III
> ... these Quaker sailors
> ... died
> When time was open-eyed,
> Wooden and childish; only bones abide
> There, in the nowhere, where their boats were tossed
> Sky-high, where mariners had fabled news
> Of IS,[a] the whited monster. What it cost
> Them is their secret. In the sperm-whale's slick
> I see the Quakers drown and hear their cry:
> 'If God himself had not been on our side,
> If God himself had not been on our side,[b]

When the Atlantic rose against us, why,
Then it had swallowed us up quick.'

IV
This is the end of the whaleroad and the whale
Who spewed Nantucket bones on the thrashed swell
And stirred the troubled waters to whirlpools
To send the Pequod*c* packing off to hell:
This is the end of them, three-quarters fools,
Snatching at straws to sail
Seaward and seaward on the turntail whale,
Spouting out blood and water as it rolls,
Sick as a dog to these Atlantic shoals:
*Clamavimus,*d* O depths. Let the sea-gulls wail

For water, for the deep where the high tide
Mutters to its hurt self, mutters and ebbs.
Waves wallow in their wash, go out and out,
Leave only the death-rattle of the crabs,
The beach increasing, its enormous snout
Sucking the ocean's side.
This is the end of running on the waves;
We are poured out like water.*e* Who will dance
The mast-lashed*f* master of Leviathans
Up from this field of Quakers in their unstoned graves?

V
When the whale's viscera go and the roll
Of its corruption overruns this world
Beyond tree-swept Nantucket and Woods Hole
And Martha's Vineyard, Sailor, will your sword
Whistle and fall and sink into the fat?
In the great ash-pit of Jehoshaphat*g*
The bones cry for the blood of the white whale,
The fat flukes arch and whack about its ears,
The death-lance churns into the sanctuary,*h* tears
The gun-blue swingle,*i* heaving like a flail,

And hacks the coiling life out: it works and drags
And rips the sperm-whale's midriff into rags,
Gobbets of blubber spill to wind and weather,
Sailor, and gulls go round the stoven timbers
Where the morning stars sing out together[j]
And thunder shakes the white surf and dismembers
The red flag[k] hammered in the mast-head. Hide
Our steel, Jonas Messias[l], in Thy side.
...

VII
...

Atlantic,[m] you are fouled with the blue sailors,
Sea-monsters, upward angel, downward fish:
Unmarried and corroding, spare of flesh
Mart once of supercilious, wing'd clippers,
Atlantic, where your bell-trap guts its spoil
You could cut the brackish winds with a knife
Here in Nantucket, and cast up the time
When the Lord God formed man from the sea's slime
And breathed into his face the breath of life,[n]
And blue-lung'd combers[o] lumbered to the kill.
The Lord survives the rainbow of His will.[p]

(Notes *a* to *o* for this poem are on pages 96–98.)

This poem, then, profoundly expresses the conviction that divine retribution and the corrupting power of human sin are underlying causes of death at sea, even as this functions as a way of making sense of such a loss. Humanity is so greedy and arrogant, so brutal, yet also so vulnerable and powerless before the sea, something that is seen above all in whaling. Yet although the ocean is a mighty, destructive force against which sailors and their craft are powerless, it too is corruptible and in turn will corrupt through the rotting flesh, the tangible material of human sin, that it carries. The whale here is a quasi-divine figure, a holy place or sanctuary, speared in its side like Christ himself, though it is not clear that it brings redemption rather than judgement alone. The hypocrisy of assuming that 'God is for us' though we work against his creatures, slaying even in the place where creation was celebrated by heaven, is expressed despite the sorrow for human loss at sea that the poem communicates. The epigraph, 'Let man have dominion over the fishes of the sea and the fowls of the air and the beasts of the whole earth, and every creeping creature that moveth upon the earth' sets whaling within the context of the deep necessity and divine injunction to exercise power over the creatures responsibly and protectively. At the same time, it suggests that death at sea and the wider pollution of the earth might be interpreted as a direct result of failure in this area, even while the whole notion of human dominion is problematized and further limited by the manifest lack of capacity by sailors to save even themselves at sea. The lowering of the prevalent yet illusory over-estimation of human powers is effected further by the idea that the people too were created with other species out of sea-slime and that dead sailors rot with other beings, lower and higher than themselves, in the water.

Key message

Human beings inevitably have a greater impact on the earth and on other life than any other species and, at times, human activity can seem god-like in its scale. Nonetheless, our common origin with other animals and the need to exercise beneficent responsibility should govern our interactions with the rest of the natural world. There are in addition good grounds for arguing that the Old Testament sense of human finitude and vulnerability before creation is one that urgently needs to be recaptured. This is especially the case where we have tampered with what Job refers to as 'what I did not understand, things too wonderful for me which I did not know' (Job 42.3).

Finally, the biblical recognition of the finitude of fish resources and of the gifting of fish is prescient for the need to limit human exploitation after generations of misguidedly presuming on the limitlessness of the sea. Fresh thought should therefore be given to the scale of fishing and to the welfare of marine life. The big challenge is whether it is possible, after centuries of scientific and technological development, to recapture a 'primitive' worldview – a worldview that dispenses with the illusions of control and allows us to recognize vulnerabilities that have long been denied but that have been as much exacerbated as mitigated by centuries of human endeavour. Having now attained an unhealthy level of power over the natural world, the challenge for humanity in the modern era is to recognize the appropriate limits of the human place within the created order and the very real responsibilities that must accompany that position.

Challenge

As always, the challenge we face as humanity is how to live in a way that cares for God's creation. It is perhaps simplistic to account for the maltreatment of the earth in terms of human greed, the desire for commercial gain, or governments wanting to demonstrate their (god-like) power to change the natural world and thereby enhance their status as 'top' nations. Undoubtedly some of these motivations apply. Sometimes, though, it may simply be that we fail fully to understand the consequences of our actions, seeing the earth (or the oceans) as so huge and our ability to inflict damage on it as so limited. While that notion may have been true 200 years ago it is no longer so.

Reflection and discussion

In our examination of Genesis 1.26, 28, we saw that it reflected a concern with survival – with the need to be safe and secure, to live peaceably on the land, with healthy progeny and sufficient food. However, the idea of 'filling' the earth, for 'dominion' and 'subjugation', suggests a wish for 'progress', for extension of one's land, increase in population, improvements in farming, and more control over one's life and the world around. If this concerns the initial growth of humanity from the point of creation until a stable, sustainable population and territorial reach was established, then it may already be understood as having reached fulfilment in Exodus 1.7, in the statement, 'But the Israelites were fruitful and prolific; they multiplied and grew exceedingly strong, so that the land was filled with them.'

A renewed attainment of this simple goal is also envisaged for those returning from Exile in Jeremiah 23.3 ('Then I myself will gather the remnant of my flock out of all the lands where I have driven them, and I will bring them back to their fold, and they shall be fruitful and multiply'). Here too blessing, prosperity and progeny are anticipated, providing a fulfilment of the Genesis promises. However, when can we ever say we now have enough, that striving and aspiration and desire for improvement has its beneficial limits? Inter-species competition, motivating the Bronze Age farmer to defend his flocks against predators, and even to seek to eliminate those creaturely rivals, has been surpassed by much-vaunted human competitiveness, whether this operates on an individual, organizational and commercial or international scale. As long as we measure our success in terms of having more possessions, or greater profits, or more power than our perceived rivals, we distort the intentions of Genesis 1.

More than that, we betray the wisdom of Jesus that true greatness comes in putting oneself last, not first. It is found in humbling oneself to serve, not to be served, and to be prepared to sacrifice oneself for others. If living in the image of God might entail service and willingness to engage in self-sacrificial giving (as the Gospels in particular suggest that it does), then measuring ourselves on the basis of our car or our house or our wardrobe, or on having more or better electronic gadgets than our neighbour, is indicative of completely having lost sight of what Christian living might entail. If Jesus called his followers to pick up their cross and follow him, to travel simply when spreading his word, and to divest themselves of worldly concerns, what do we think we are doing if our favoured recreational activity is shopping, and our talk is of our latest home improvement, the qualities of our car, and which products we should aspire to next?

Genesis 1.26, 28 encapsulates a programme for a thriving humanity that can make its way in the world, but not for excessive exploitation of the earth and rampant consumerism. Many passages concerned with fishing reveal a real revulsion at excessive greed and unrestrained consumption, as well as a sense of this food as a gift from God. Not only have we collectively ignored this, but ironically our greed will deplete not only every resource on which we depend, as well as causing unprecedented levels of extinction among other species, but it will ultimately make our lives as we know them unsustainable.

Genesis 1 begins with a world that was 'very good'. Scientific understandings of the world reveal just how astonishingly good it is, finely tuned to an improbable degree in order to permit life – wonderful, diverse, teeming and plentiful life.[30] It contains everything necessary to support living things, in an amazingly sustainable form. However, the earth is (by and large) also a closed system in terms of the materials necessary for life, with the exception of the energy input from the sun.

This has been visualized as 'Spaceship Earth':[31] just as on a spaceship, the only resources available to you are those with which you commenced your journey, so with the earth we have only what is here, and no chance to 'restock' from elsewhere. We are depleting every imaginable resource, melting the icecaps, releasing ever more carbon into the atmosphere, and pumping aquifers dry. The scale of our overconsumption is made especially apparent on the many websites detailing the amount of waste we each produce over a year or in our lifetimes. Possibly even more striking is the contrast between the huge amount of waste created in the affluent West as compared to sometimes very minimal discards of more traditional societies. Alarm is expressed at the consequences of the developing world 'catching up' with Western consumerism. In reality, though, it is the West that needs to rebalance and reprioritize in order to live a life of more sustainable consumption.

We are blessed with the means not only to survive but also to flourish. However, if we are to live in the image of God, Jesus' model of service, rather than aspirations to conspicuous consumption as a mark of status, should inform not only our relation with other people, but also with the rest of creation. Not 'Do I want it?' and 'Can I buy it?', but 'Do I really need it?' and 'Can I live without it?' should be principles that inform our purchase of any non-consumable product. Dieting is sometimes framed in terms of changing our relationship with food: we may eat not through hunger, but through loneliness, boredom, habituation, or a subconscious need for a boost. In the same way we need to change our relationship with 'stuff', taking only what we need and enjoying a 'lighter', 'slimmer' existence that brings us closer to our true place within creation and that lessens our negative impact on other creatures.

In light of the above, here are a few thoughts to consider:

- How do you view our place in creation? Are we animals? Do we have the same value as other creatures, or are we really somehow more important? Even if you think we are superior, would this give us extra rights or extra responsibility (or both)? Would it license us to treat other creatures as of less value than us?

- What might it mean to be made in the image of God?

- How do you understand the mandate to 'fill the earth and subdue it and have dominion' over the other creatures? Does this allow a utilitarian attitude to the rest of creation and license a certain amount of exploitation? What else might it mean?

- The passages we have looked at suggest quite modest aspirations as regards what might constitute a good (or even miraculous) catch. The scale on which 'factory' fishing operates now is unimaginably greater. However, this can also provoke wider questions about lifestyle and whether we are generally too greedy and exploitative of the world around us. How much do we really need, and how many luxuries might we live without? Would a simpler lifestyle bring benefits, or only losses?

Action

Confronted with the scale of some of the problems – factory fishing, major oil pollution incidents, the drying up of an inland sea – we may feel there is little that we can do. However, that is not the case. Even small actions are worth pursuing. We can start at the personal level by buying fish to eat that have been sustainably caught, and many other small changes. Escalating our activity to a low key political level may also be appropriate, such as writing to our local MP in the UK, or equivalent if living elsewhere. A more overt political response may be the next step: joining a religious or secular campaigning organization that is trying to make changes to the status quo that is harming the earth. In fact, there is no limit to where small initial steps might take us in trying to fulfil God's mandate to care for his creation.

However, one of the most important things we can do is to change our attitude to consumption and to make it a priority to minimize what we buy or use, instead of striving for more. Choosing products like home fittings or clothes on the basis of what will last (both in terms of quality and style) instead of following the latest fashion requires a major shift in thinking for many of us. Nonetheless, if we can return to being content with less and finding gratitude in the many blessings we have already enjoyed, instead of living with habitual dissatisfaction and material aspiration, the beneficial outcomes may include not just planetary health but human wellbeing as well.

Notes

1 Rising ocean temperatures and ocean acidification are discussed in Chapter 7.

2 From the perspective of astronauts heading for the moon, at least. The phrase 'The Blue Marble' is the designation of a photograph taken on 7 December 1972 from the *Apollo 17* spacecraft at a distance of about 28,000 miles from earth. It is an iconic image and has been widely distributed (possibly one of the most widely distributed images in history). You can find it on the Wikipedia page 'The Blue Marble' or at www.nasa.gov/content/blue-marble-image-of-the-earth-from-apollo-17.

3 El Niño and its consequences are discussed further in Chapter 6.

4 'Spill' is something of an understatement!

5 See Chapter 7.

6 The largest was – and still is – the Caspian Sea, and the second and third were Lake Superior and Lake Victoria.

7 The Dead Sea – the saltiest sea – is about three times as salty as the Aral Sea.

8 Some graphic images of lake reduction are available at mnn.com

(Mother Nature Network), and links to information on various lakes that are drying up can be found through the Wikipedia page, 'List of Drying Lakes'.

9 In Revelation, although a new heaven and earth are mentioned in 21.1, the new creation seems to be focused only on the sacred, urban space of the sanctuary-like holy city, the New Jerusalem. Despite the mention of the waters of 'the river of the water of life' streaming out, it simply flows 'through the middle of the street of the city' (22.1), and there is little interest in any space beyond this. This new creation, then, is confined to holy space, and there is an intense interest in the exclusion of anything that might compromise this: nothing unclean will enter it, but only those whose names are entered in the Book of Life (21.27); there will be no death or mourning or crying or pain (21.4); no night (21.25); no need for sun or moon because of its constant light (21.23, 22.4); no temple even (21.22), because of the divine presence – in other words, no scope for anything that might be open to corruption. It is not surprising, then, that the other primordial element besides the night/darkness, namely the sea/deep, will also be absent (21.1), even though the presence of life-giving water (21.6, 22.1–2; also associated with the deep and represented in the temple) will be an important feature.

10 To explore these issues further, see Tom Wright, 2007, *Surprised by Hope*, London: SPCK, and Christopher J. H. Wright, 2010, *The Mission of God's People*, Grand Rapids: Zondervan.

11 One might compare the biblical perspective on the over-harvesting of the manna provided by God in the wilderness, in which lack of trust in God is also an important motif. Note especially the idea that however much or little people gathered, they each had as much as they needed, neither more nor less (Exodus 16.16–18), while those who tried to store more excited Moses' anger (vv. 19–20). The striking down with a plague of those who craved meat in the wilderness provides another illustration of the same values (Numbers 11), while the condemnation of greed (in this case, the excessive storing up of crops) is expressed especially powerfully in the New Testament in Luke 12.14–21.

12 Matthew 4.19; Mark 1.17; similarly in Luke 5.10, following the miraculous catch of fish, Jesus says, 'Do not be afraid; from now on you will be catching people'. (The word used here in Luke means specifically to 'catch alive' rather than simply to 'fish', although the context still makes clear the fishing analogy.)

13 Luke 5.1–11.

14 John 21.1–14.

15 See the feeding of the 5,000 (Matthew 14.19; Mark 6.41; Luke 9.16; John 6.11), the feeding of the 4,000 (Matthew 15.36; Mark 8.6–7), and John 21.13, where after the miraculous catch of fish the resurrected Jesus 'came and took the bread and gave it to them, and so with the fish'.

16 In the Gospels he is said to take a cup and give thanks in the last supper (Matthew 26.27; Mark 14.23; Luke 22.17; and probably also implicitly in Luke 22.20). 1 Corinthians 10.16 mentions 'the cup of bless-ing that we bless', and of course 1 Corinthians 11.25 also recollects the sharing of the cup at the last supper.

17 It seems especially from Luke 24.30–31, 35 that this was a distinc-tive action of Jesus. However, as these gestures apply also to the blessing, breaking and distribution of fish, and this, together with bread, was the staple food of Galilee, it may be that the action itself, rather than bread as the object, is the distinguishing feature. The description as we have it was probably also influenced by eucharistic practice.

18 Examples are the fresco of the *Fractio Panis* (breaking of the bread) in the *Capella Greca* (Greek Chapel) in the Catacomb of St Priscilla on the Via Salaria Nova in Rome, which may be early second century, and the crypt of Lucina in the catacomb of St Callistus. Textual allusions to fish in association with the bread and wine occur in the late second-century Inscription of Abercius, Bishop of Hierapolis in Phrygia, and in that of Pectorius of Autun, which may date to the end of the second or early third century, though dates as late as the sixth century have also been proposed.

19 The boy is mentioned only in John's Gospel (6.9).

20 Matthew 14.13–21; Mark 6.30–44; Luke 9.12–17; John 6.1–14. The point that the 5,000 men did not include women and children is made explicit only in Matthew (14.21).

21 Matthew 15.32–38; Mark 8.1–9.

22 Matthew 15.32; Mark 8.2–3.

23 www.fao.org/3/a-i3807e.pdf. This compares with about 90 million tonnes of aquaculture production (including about 24 million tonnes of aquatic plants), a figure that is growing. Sardine catches alone amounted to over a million tonnes in the same year (www.fao.org/fishery/species/2910/ en).

24 Numbers 11.21. The very large quantity in Numbers has been the cause of much debate. The different ways of understanding them are summarized in detail in R. Dennis Cole, 2000, *Numbers*, New American Commentary 3B, Nashville, TN: Broadman and Holman, pp. 78–82. This includes the possibility that the word usually translated 'a thousand' may denote something else, such as a 'clan' or a military troop or division, or that the numbers are symbolic or purposefully hyperbolic, or that they represent some form of gematria (in which numerical values are assigned to words or phrases and understood to have special significance), or that they may perhaps relate to population numbers at a subsequent time.

25 The NRSV reads here, '"Are there enough flocks and herds to slaugh-ter for them? Are there enough fish in the sea to catch for them?"', but it does not accurately communicate the passive force of the Hebrew verbs.

26 This question invites in the modern reader the further matter of

what constitutes 'sufficient', since sufficiency to be fished in a sustainable way must be distinguished from the absolute quantity of fish that could in principle be drawn on without regard for the numbers required to maintain stock levels or to avoid having an adverse effect on other organisms within the same food web. From this perspective, Moses' words might be recognized as, far from reflecting the ignorance of the ancients with regard to the capacity of the sea to nurture vast fish reserves, fortuitously prescient and insightful (if only accidentally so) in recognizing the limitations of what it can provide. It may indeed be the case that if 600,000 people (not to mention the additional numbers created by the women, children and even non-fighting men possibly not included here) feasted for a month on fish drawn from the same section of coast this would ultimately deplete fish stocks and diminish their capacity sustainably to replenish themselves for some time.

27 In addition, Jeremiah 16.16 provides an example where the people who will go into hiding are warned that they shall be fished out and hunted, for Yahweh shall 'send for many fishermen and they shall fish for them', and he shall 'send many hunters and they shall hunt them from every mountain and from every hill and out of the clefts of the rocks'. The terror of being fished or hunted and the merciless thoroughness of the fisherman or hunter is forcefully conveyed in this passage.

28 This misguided and ugly greed is caricatured in the picture of the foe sacrificing to his net since he has gained so much by it and evidently seeks for more.

29 The idea that the fish are vulnerable because they lack a *mōshēl*, a ruler who may lead and protect them, creates a resonance with a further passage, Psalm 8.6, in which the same verbal root *m-sh-l* recurs. It reads 'You have given them (that is, human beings) dominion over the works of your hands' – that is, you have made them rulers over the works of your hands. The verb *m-sh-l*, used in each of these contexts, is employed here in the sense of ruling over one's fellow-countrymen and is not therefore used of subjugation, domination and oppression, although certainly there can be good rulers and bad. But essentially to 'rule over' entails responsibilities towards the people, and benefits for them in return. These passages, Habakkuk 1 and Psalm 8, taken in combination, point both to a need (the vulnerability of fish before humanity) and its resolution (stewardship by a humanity that recognizes itself as one creature among other fellow-creatures and with responsibility to ensure the welfare of all). The fish have no 'ruler' to protect them, but if we are commissioned to 'rule' over God's creation, we should arguably be fulfilling that role.

30 'Fine tuning' and related cosmological questions are examined from a Christian perspective in Rodney D. Holder's *Big Bang, Big God: A Universe Designed for Life?*, Oxford: Lion Hudson, 2013.

31 See Kenneth E. Boulding, 1966, 'The economics of the coming

Spaceship Earth', in H. Jarrett (ed.), *Environmental Quality in a Growing Economy*, Baltimore, MD: Resources for the Future/Johns Hopkins University Press, pp. 3–14.

Notes on the poem, 'The Quaker Graveyard in Nantucket. For Warren Winslow, Dead at Sea' by Robert Lowell

a 'IS' subtly identifies the whale with Christ, since it alludes to Gerard Manley Hopkins's line 'What Christ is [...] IS immortal diamond' ('That Nature is a Heraclitean Fire') and ultimately to God's self-revelation as 'I am' in Exodus 3.14.

b Here the allusion is to the opening lines of Psalm 124:

¹If it had not been the LORD who was on our side
   – let Israel now say –
²if it had not been the LORD who was on our side,
   when our enemies attacked us,
³then they would have swallowed us up alive,
   when their anger was kindled against us;
⁴then the flood would have swept us away,
   the torrent would have gone over us;
⁵then over us would have gone
   the raging waters.

Ironically, of course, the whale has just been identified as Christ; the whalers themselves are the attackers; and although they imagine God to be on their side, they perish in the sea – exactly the opposite outcome to the salvation celebrated in the psalm. The Quaker whalers' comfortable assumption of divine approval and protection is, then, directly challenged here.

c The *Pequod* is the whaling boat in *Moby Dick*: the disapproval of Lowell is eloquently communicated by the perception that it will be sent 'packing off to hell'.

d *Clamavimus* (Latin, 'We have cried') refers to Psalm 130.1, 'Out of the depths I cry to you, O LORD'. We know though that this prayer will remain unanswered, and are left wondering if the whalers are not just petitioning God but perhaps also on some level the whale.

e 'Poured out like water' again picks up on biblical imagery, but in a way that operates on different levels. This phrase is drawn from the lament Psalm 22, as the psalmist begs God to attend to his suffering. There is extra irony, however, that in the New Testament this psalm is understood as being fulfilled by Jesus on the cross (in this case, when mixed blood and water poured from his side when he was pierced, John 19.34). Most often in the Bible, however, this phrase occurs in respect of the command that the blood of sacrificial animals should be 'poured out like water' that all may go well for the slaughterers. Again, the dying whale seems to be as much in view as the ostensibly pious sailors.

f Odysseus was lashed to the mast to prevent him from throwing himself into the waves after the Sirens lured him with their song. Here, it speaks of the whaler's morbid rigidity, but may hint also at his desire to go into the sea after the whale. In *Moby Dick*, the captain, Ahab, was bound to the white whale after spearing it.

g The Valley of Jehoshaphat is a place of judgement in Joel 3.2, 12, but there is an extra layer of meaning insofar as the good Judean king Jehoshaphat went up to battle with the notoriously wicked Israelite king Ahab, where the latter was killed as an act of divine judgement. Since Ahab is the whaling master in *Moby Dick*, the choice of this name can hardly be coincidental.

h In one of the most powerful allusions in the poem, the whale's body becomes 'the sanctuary', the holy place and dwelling-place of God, where he is encountered and worshipped. We are reminded that Jesus, according to John 2.19–22, offered the Jews a sign that if they destroyed the temple, he could resurrect it in three days, thereby referring to 'the temple of his body'. The whale seems to take on a similar role here.

i 'Swingle' here seems to describe the whale's agonized writhing as it dies, its flukes (tail lobes) about its ears.

j Job 38.8 joyfully presents creation as that time

when the morning stars sang together
　and all the heavenly beings shouted for joy.

What a ghastly irony, then, that the whale should die just there.

k In *Moby Dick*, a sailor fixed a red flag to the mast as a final attempt to attract help as it sank. But perhaps the red, too, reminds the reader of the blood shed on the boat by its crew.

l 'Jonas' or 'Jonah' is, of course, well-known as he who was inside the whale, but here his name is combined with that of the Messiah. Jesus' side was pierced on the cross (John 19.34, echoing Psalm 22.14), so again the 'sanctuary' of the whale's body seems to be analogous to the cruci-fied Christ. The Jonah–Messiah allusion goes a little deeper, however, as Jonah's three days in the belly of the whale is in the New Testament inter-preted as a 'type' prefiguring Jesus' three days in the belly of the earth before his resurrection (Matthew 12.39–41). The force of this comparison, though, is not one of resurrection and hope, but as a sign of judgement: the people of this generation are so wicked that 'The people of Nineveh will rise up at the judgement with this generation and condemn it, because they repented at the proclamation of Jonah, and see, something greater than Jonah is here!' (Matthew 12.41; see also Matthew 16.4 and Luke 11.29–30, 32).

m The poem draws towards its close with a picture of the Atlantic which, though once the market ('mart') for trading ships ('clippers'), is polluted with the corpses of dead ('blue') sailors, sea monsters and fish. Milton in *Paradise Lost* depicts sea monsters and fallen angels (Satanic powers) as being in the sea, but here the whales are not evil monsters, and the angel is not a fallen one but 'upward'. Human sin seems to have polluted all.

n God breathing the breath of life into humanity is a familiar image drawn from Genesis, but here humanity is envisaged as coming from the sea. The destiny of these drowned sailors is not 'dust to dust' but 'sea slime to sea slime'. Various creation stories portray life as emerging from the water rather than the soil but, as employed here, it may indicate that we are ultimately part of the sea, or indeed born for destruction in a hostile environment.

o It is not clear what the 'blue-lung'd combers' are – waves, whales or the dead whalers themselves? But their violence, and the ultimate defence-lessness of humanity, is undeniable.

p 'The rainbow of His will' alludes to the Flood story, at the end of which God places a rainbow in the sky as a sign of his promise never to flood the earth again. It seems somewhat ironic to refer to this promise in the context of a poem about judgement and drowning, but it affirms the persistence of God and his sovereign independence to do his will.

5

The Sacred Sea

Holy water?

A striking aspect of the sea, both from a scientific and theological perspective, is its otherness: its depths are inaccessible and still, to a considerable degree, unknowable to human beings. The Bible, however, recognizes that at a more profound level the sea and its 'deeps' (understood as great reserves of water feeding the sea) are a sacred space known only to God. One place where this emerges particularly clearly is in the so-called first divine speech in the book of Job. Job is depicted as a righteous man who is nevertheless subject to the most awful suffering, losing his children, his possessions and his health, at the end of which he seeks vindication from God. In the final chapters of the book, Yahweh emerges out of the whirlwind and challenges Job: Where were you when I created the world? Can you lead the stars out or manage the weather? Do you look after the wild animals and provide for their needs? Who are you to question me when you know so little? As part of this challenge, Yahweh highlights the major elements of creation of which Job knows nothing:

> [16]'Have you entered into the springs of the sea,
>   or walked in the recesses of the deep?
> [17]Have the gates of death been revealed to you,
>   or have you seen the gates of deep darkness?
> [18]Have you comprehended the expanse of the earth?
>   Declare, if you know all this.' (Job 38.16–18)

The speech continues by God asking Job where light and dark, snow, hail and wind are stored (presumably somewhere in the heavens or under the earth). Of course Job has no idea about these things, for the depths of the sea, the heavens and the underworld, or even understanding the full expanse of the earth, are beyond him. Such knowledge is God's alone, and such places – both the heavens and the depths beneath – must remain entirely out of Job's reach and experience.

Job ultimately responds to God's answer out of the whirlwind with shame and humility, acceding that 'I have uttered what I did not understand, things too wonderful for me, which I did not know ... therefore I despise myself, and repent in dust and ashes' (Job 42.3, 6). The nineteenth-century American poet Walt Whitman seems to express a similar impulse in his poem 'As I Ebb'd with the Ocean of Life', prompted again by reflection on the ocean:

> As the ocean so mysterious rolls toward me closer and closer,
> I too but signify at the utmost a little wash'd-up drift,
> A few sands and dead leaves to gather,
> Gather, and merge myself as part of the sands and drift ...

> I perceive that I have not really understood any thing,
>   not a single object, and that no man ever can,
> Nature here in sight of the sea taking advantage of me
>   to dart upon me and sting me,
> Because I have dared to open my mouth to sing at all.

However, the extent of human ignorance and weakness is only one aspect of the lesson of Job. The connection of the sea with the heavens and even with the world beneath in the passage we have just looked at is no coincidence because we encounter it too in the theology of the temple. In fact, according to the description of the construction of the temple and its furnishings under Solomon (outlined in 1 Kings 7.13–14, 23–26, and in the parallel account in 2 Chronicles 4.2–5), an immense bronze 'sea' was actually situated in the temple itself. At 5 cubits high (a little over 7 feet, or about 2.2 metres according to a common estimate) and twice as wide, its dimensions would have rendered regular use extremely difficult, but we can be sure that it would have had symbolic significance. Its capacity, according to Kings, was 2,000 baths (perhaps 10,000 gallons, or about 44,000 litres, though this is debated),[1] while Chronicles claims that it held 3,000 baths (15,000 gallons, or 66,000 litres, according to some estimates), so it would have far exceeded what might have been required for purification purposes alone.

However, symbolically, the idea of the sea as being under divine control was of critical importance. The sea was from this perspective a sacred space, an essential aspect of God's dwelling-place and equally vital as a source of life. Such ideas may in certain respects seem remote from the modern world. Nonetheless, they also have a deeper resonance, as the following section of Byron's lengthy narrative poem, *Childe Harold's Pilgrimage*, shows:

> Roll on, thou deep and dark blue Ocean – roll!
> …
> Thou glorious mirror, where the Almighty's form
> Glasses itself in tempests; in all time,
> Calm or convulsed – in breeze, or gale, or storm,
> Icing the pole, or in the torrid clime
> Dark-heaving; boundless, endless, and sublime –
> The image of Eternity – the throne
> Of the Invisible; even from out thy slime
> The monsters of the deep are made; each zone
> Obeys thee; thou goest forth, dread, fathomless, alone.

The idea of the sea as 'the throne of the Invisible' beautifully encapsulates an aspect of ancient temple theology that we shall return to shortly. However, the mystery, inhospitability and unknowability of the sea is not a theme that is confined to poetry or to the Bible, but is also reflected in scientific study of this little-understood and extraordinary part of our planet.

Twenty Thousand Leagues Under the Sea

The title of Jules Verne's famous book is often taken to refer to the depth of the ocean, but it actually describes the distance that

Captain Nemo's submarine *Nautilus* travels in the ocean. This popular misunderstanding in itself reminds us how slender our knowledge of the depths actually is and how open we are to wonder at this extraordinary and mysterious space. In Verne's book a league was equivalent to 4 kilometres, which is approximately the average depth of the world's ocean. Since Verne's time, submarines capable of such long voyages have actually been built – nuclear submarines that can stay submerged for months and travel long distances underwater. Nevertheless, most modern submarines are still confined to the top few hundred metres of the ocean because of the tremendous pressures that the water above exerts on their hulls. (Every increase in depth of 10 metres adds the equivalent of one atmospheric pressure.) To go deeper requires specialized deep submersibles, specifically designed to withstand the huge pressures at great ocean depths. It is only recently that a few human beings have ventured to the ocean deeps in such vessels, and much of the ocean floor still remains unmapped and unexplored. Our experience of these regions remains very limited and vast tracts of the deep ocean have never been seen by human eyes.

However, though remote to us, the deep ocean can be an extraordinarily dynamic place, where the earth's crust continually renews itself. Here, in the Mid Pacific Ridge and Mid Atlantic Ridge, the movement of tectonic plates as they spread apart creates a volcanically active zone. Volcanic activity can even create islands in the sea where people may eventually live, such as the Hawaiian Islands, one of which, The Big Island, still has an active volcano, Mauna Loa.[2] The continental rift in Iceland is the continuation of the spreading oceanic Mid Atlantic Ridge and is one of the few places on earth where a mid-ocean ridge can be seen on land. There, two parts of the island are moving apart with an average speed of 2.5 centimetres per year, but it is a strange experience to be aware of this in such a tangible way on land.

The counterpart of such regenerative processes is subduction, the movement of one plate under another at the edge of the ocean, which creates huge ocean trenches. One such subduction zone is the Mariana Trench in the western Pacific, to the east of the Mariana Islands in Micronesia. Here one tectonic plate, the Pacific plate, is thrust under another, the Mariana Plate. This provides the location for the deepest known point in the ocean, the Challenger Deep, found at the southern end of the Mariana Trench. This is almost 11 kilometres deep (measured to be 10,994 metres), which far exceeds, in distance below sea level, the height of Mount Everest above sea level (measured to be 8,848 metres). The Challenger Deep was first visited by humans, using a deep sea submersible, in 1960, and more recently in 2012 by the *Titanic* film director James Cameron. Essentially the depths of the ocean continue to be a little-explored and mysterious part of the earth.

The movement, growth and destruction of tectonic plates at the bottom of the deep ocean is part of the dynamic earth, with crustal material being created and spreading from mid-ocean ridges and being subducted back into the earth's interior in regions such as the Mariana Trench in the Pacific. Over hundreds of millions of years, unobserved by humans, these processes have created continents and mountains and then repositioned them on the face of the earth. At times in the earth's history all the landmasses of the world have been connected in a single continent as the tectonic plates moved them around. This last happened about 300 million years ago and the resulting supercontinent is known as Pangaea. This supercontinent existed until about 175 million years ago when it began to break up as the tectonic plates continued to move. Pangaea was totally surrounded by the sea, a super-ocean called Panthalassa.[3] While the configuration of the world's continents has changed, the presence of water on the earth has been ubiquitous, whether as water vapour in the atmosphere, or water in the oceans, or ice on land and sea.

Water, water, everywhere

Importantly, the fact that the earth is just the right distance from the sun to support water in all three phases – gaseous (water vapour), liquid (water) and solid (ice) – is critical to the way the climate system works and to the development of life on earth. However, the origin of water on our planet is an area of active research with, as yet, no scientific consensus regarding the answer.

There are two main competing theories. Since comets contain a high percentage of water, one possibility is that cometary bombardment early in the life of the earth delivered water to the planet and that this was sufficient for the oceans to come into being. The other dominant theory is that when the earth took shape roughly 4.5 billion years ago, the material from which it formed contained water, which would have been released into the atmosphere through tectonic activity like volcanoes. Again this would have happened early in the life of the planet when volcanic activity was much greater than it is today. Of course, it is possible that both processes played a role, each contributing to the existence of the oceans. Whichever is the case, the important point is that water was present from an early stage in the life of the earth and was critical both to how the planet developed and ultimately to how life originated and evolved. The oceans presently cover 71 per cent of the earth's surface and contain 97 per cent of the world's water, so remain its dominant feature, as pictures of our planet as viewed from space so strikingly show.

It is the presence of water and tectonic processes that have largely shaped the earth's surface. Tectonic processes, such as the formation, movement and subduction of the tectonic plates, and the earthquakes and volcanoes associated with this, have created and rearranged the land masses and continents. As a result, the shape of the oceans today differs significantly from those that existed hundreds of millions of years ago. Likewise, the cycle of water evaporating from the ocean, being moved by the atmosphere and falling on land and running in rivers back into the sea, has shaped the earth through erosion. In fact, the seas became salty through this process over many millions of years as the rivers washed minerals from the land into the ocean. In colder phases of the earth's existence, when large amounts of water have been trapped in glaciers, these have moved across the earth's surface, gouging out and smoothing out valleys and other landscape features.

Not only has the presence of water affected the surface of the earth, but the seas have been an integral part of the planetary system over eons, sustaining life and providing an environment in which it can flourish. It may even have provided the conditions in which life began.

Water – the fount of life

Life as we know it would not be possible without the presence of water, and indeed all living things contain significant amounts of water. About 60 per cent of an adult human body is water (though this varies somewhat with age, gender and body type), and we need 2 to 3 litres of water a day to survive healthily. Some organs like the brain contain a higher percentage of water (73 per cent) while other parts of the body – like bones – have a lower percentage (31 per cent). Similarly, there is considerable variation in the proportion of water comprising different organisms. For example, jellyfish might be composed of as much as 95 per cent water.

Water is a most unusual substance and it is its peculiar properties that enable it to sustain life. A particularly distinctive feature is that its maximum density occurs at about 4°C, so as it cools to freezing point it becomes less dense. This too is absolutely critical to life on earth, since it means that lakes and seas freeze from the top and the ice floats on the surface. Were it not for this property, they would freeze from the bottom up, potentially destroying all life within them. In addition, the floating ice acts to insulate the water beneath and allows it to stay liquid, so aquatic creatures are able to survive.

Although most of the world's water is in the ocean, the part that is cycled through the atmosphere is absolutely critical to life too. In the water cycle, evaporation from the oceans and transpiration from plants on land release water vapour into the atmosphere. It is then transported round the globe and, when cooled, condenses and falls as rain, before returning to the oceans via rivers. Some of this water (primarily from the oceans but also from the land) then evaporates again, thus continuing the cycle. This is a vital part of the global climate system.

The Bible, too, is acutely aware of the central place of the ocean in the water cycle. However, although it knows that water rises to form clouds, it understands the principal mechanism in the cycle as subterranean: water was believed to flow underground back from the sea to reservoirs under the earth, and probably especially in the mountains. These reservoirs were, like the sea, known as 'the deep'. From them, the water would rise to the surface of the earth and flow back down again as streams. In this way, the Bible appears not to make a distinction between saline and fresh water, but to think of the deep rather as a vast reservoir of water feeding both freshwater sources and the sea.

Interestingly, although this view of the movement of water is not quite accurate, it is true that the earth's great store of water – the ocean – is salty, but when water evaporates it leaves the salt behind, as many of us know from chemistry experiments done in our schooldays. This means that the water cycle provides fresh rather than salty water to the atmosphere that then falls as rain. Rainfall is slightly acidic as it contains dissolved carbon dioxide from the atmosphere, so forming carbonic acid. The rainfall erodes rocks, and being acidic it breaks them down. It thereby releases minerals and salts that are washed away and eventually reach the sea as river run-off from the land. Organisms in the sea use some of the dissolved minerals and salts, but the rest accumulates in the water and over millions of years this process has led to the oceans becoming more saline. In addition, hydrothermal vents deep in the ocean re-cycle water through the earth's crust and so contribute to the flux of minerals and salts in the ocean. Finally, submarine volcanoes, though less common, also add minerals and salts to the oceans when they erupt below the ocean surface. All these geophysical processes, operating over millions of years, have created the ocean environment that we know today which is teeming with life of all kinds.

The deep fountains of the sea

Earlier in the book we described the black smokers that exist in the deep ocean where very hot water, rich in minerals, emerges from hydrothermal vents on the seabed. This water will have been heated geothermally from deep within the earth, so when it emerges from the vents and encounters the cold waters of the deep ocean, this precipitates the minerals. That is, the minerals in the water cool and become solid particles in what appears as a black cloud – hence the name 'black smokers'. Black smokers were discovered about 40 years ago and are generally associated with the volcanically active deep ocean ridge systems. The water emerging from them is very hot (up to 450°C) and acidic.

More recently, in the year 2000, a new type of hydrothermal vent has been discovered. This new type of vent differs in being some distance away from the volcanically active areas and the water emerging from the vent, while hot (about 200°C), is not as hot as that associated with the black smokers. More importantly, the water is alkaline rather than acidic. This type of vent is thought more closely to resemble the conditions in the early history of the earth when life emerged (about 3.5 to 3.8 billion years ago).[4]

The earth formed about 4.5 billion years ago and was initially a very inhospitable environment for life to develop on its surface, being subject to high temperatures and meteorite bombardment. Oceans came into being some 750 million years later and the deep ocean would have been a much safer and more stable environment in which for life to begin. Fossil evidence suggests that hydrothermal vents existed in the early earth's ocean. So these 'fountains of the deep',[5] with their peculiar environments existing under tremendous pressure and rich in chemicals, might possibly have provided the conditions in which the organic molecules necessary for life to emerge might have formed. They are therefore a possible location for abiogenesis – the formation of life from non-living matter. Research into this possibility continues but we do not yet know the answer (and perhaps never will).

The deep and the 'fountains of the deep' continue to fascinate and remain largely unknown, unexplored and hidden from human sight. Did life begin in the strange environment associated with hydrothermal vents? How many unusual creatures and features of the deep ocean have we yet to discover? There is something mysterious about the ocean deeps, something that evokes a sense of awe.

The seat of God in the heart of the sea

Our new understanding of the deep sea, embryonic as it is, is a far cry from the world of the Bible, where such scientific discoveries were entirely unknown. Despite this, there are striking, though surprising, resonances between these oceanographic insights and some of the beliefs surrounding the 'heart of the sea' reflected in biblical texts concerning the theology of the temple. There is obviously no direct relation between the two understandings of the sea: they are founded on different premises and developed for independent reasons, as well as being part of contrasting world-views. The temple theology we are about to explore here is also only one biblical perspective among many on the sea. Nonetheless, both worldviews, those of temple theology and deep sea biology and geology, invite us to reflect on the sea as a sacred space, one that is inaccessible to us and inextricably linked both with death and life.

Heavenly temple – earthly temple

To understand this, we need to enter the temple culture of the ancient Near East. According to this way of thinking, the distinction between God's heavenly temple and those here on earth cannot be drawn rigidly. For a start, ancient sanctuaries were founded on sites where the divine presence had already been encountered and where it continued to be made manifest. Sanctuaries were intrinsically holy places: in a sense, they were discovered (or rather, revealed), not just made on a human whim.

This comes across most strongly in the story of 'Jacob's ladder' (Genesis 28). Jacob dreamed that there was a ladder reaching up into heaven, with angels ascending and descending on it. When he awoke, he said,

> [16]"Surely the LORD is in this place – and I did not know it!' [17]And he was afraid, and said, 'How awesome is this place! This is none other than the house of God, and this is the gate of heaven.'
> [18]So Jacob rose early in the morning, and he took the stone

that he had put under his head and set it up for a pillar and poured oil on the top of it. [19]He called that place Bethel; but the name of the city was Luz at the first. (Genesis 28.16–19)

The sanctuary at Bethel (which literally means 'house of God') offered a unique point of encounter with the true divine reality – a reality that was not normally humanly accessible, but that had been revealed to Jacob. It marked a spot where God already was. Another reflection of the same idea was the conviction that the temple was built according to God's instruction and plan:[6] it was where (and how) God himself had chosen to dwell.

So temples themselves replicated – and, more than that, actually embodied and made present – God's heavenly abode. The powerful sense that the temple in Jerusalem merged into the reality of the heavenly temple is seen clearly in Isaiah's call vision (Isaiah 6), which begins as follows:

[1]In the year that King Uzziah died, I saw the Lord sitting on a throne, high and lofty; and the hem of his robe filled the temple. [2]Seraphs were in attendance above him; each had six wings: with two they covered their faces, and with two they covered their feet, and with two they flew. [3]And one called to another and said:

'Holy, holy, holy is the LORD of hosts;
the whole earth is full of his glory.'

[4]The pivots on the thresholds shook at the voices of those who called, and the house filled with smoke. [5]And I said: 'Woe is me! I am lost, for I am a man of unclean lips, and I live among a people of unclean lips; yet my eyes have seen the King, the LORD of hosts!' (Isaiah 6.1–5)

The vision seems to be set in the Jerusalem temple (Isaiah 6.2, 4), yet we are told that it was filled by the hem of God's robe, while its door pivots shook at the divine presence and the building filled with smoke. Even more suggestive of the heavenly temple, Isaiah saw 'the Lord sitting on a throne, high and lofty', surrounded by seraphim and addressing his council (as if in heaven). Only priests could enter the inner part of the temple, and high priests alone could venture into the holy of holies, so it is often thought that Isaiah himself must have been a priest. However, the vision also reflects a belief that by entering this highly sacred area, the priest actually came into the presence of God and in this heavenly world had access to the secrets of creation and history determined in the divine court. The coalescence of the earthly and heavenly temples is also apparent in Psalm 29, for in v. 1 the 'heavenly beings' are called upon to glorify and worship Yahweh, but by v. 9 it is said that 'in his temple all say, "Glory!"' The praise of the people and the host of heaven thus merge into a single whole.

The temple may in fact be understood as representing the intersection, the meeting point, of heaven and earth or, as Richard Clifford puts it, 'the point where the earth touches the divine sphere'.[7] This idea was expressed visually in the Babylonian ziggurats, the famous stepped temple towers reaching up towards heaven like layers on a square wedding-cake with a shrine at the top.[8] They are described in ancient Mesopotamian texts by phrases like 'Bond of Heaven and Earth' or as 'like a mountain in heaven and earth which raises its head to heaven'.[9]

The living waters of the temple

But the temple was not just the meeting place of heaven and earth. God is God of the whole cosmos, and Lord of life and death as well as of heaven and earth, so the temple too was the place where all these spheres uniquely intersected – including the regions under the earth as well as above. The temple was also perceived as a source of life. Ancient texts (dating to about 1200 BC) from ancient Ugarit in Syria describe the dwelling-place of El, the Canaanite high god,[10] as a mountain 'at the source of the two rivers, at the midst of the spring of the two deeps'. It was understood as both reaching up to heaven (hence being a high mountain) and down to the depths (hence being in the midst of the spring of the two deeps), and as the source of life-giving waters.

A similar idea is apparently present too in respect of Jerusalem, where there is 'a river whose streams make glad the city of God' (Psalm 46.4); or, as Psalm 36 puts it, 'all people ... feast on the abundance of your house, and you give them drink from the river of your delights. For with you is the fountain of life' (Psalm 36.8–9). The idea of joy, delight and abundance, and the flowing forth of the river from the holy sphere, was a corollary of God's lordship over all that is, and in particular of his dispensing of life and blessing from his throne and temple, as symbolized by the river.

The idea of God as providing nourishing stream water is mirrored in the frequency of shrines from many periods and cultures that are placed at water sources, and it finds particularly striking expression in Egypt at the temple of Karnak, where there is a large sacred lake (120 metres × 77 metres) lined with stone. However, it is also an important theme in the Bible, not least in the furniture of the Jerusalem temple itself, since each aspect of its design was highly symbolic. For example, the temple is described as having been decorated with lilies and pomegranates, palm trees, gourds and open flowers (1 Kings 6.32, 35, 7.18–20, 22, 36, 42), besides incorporating 'the House of the Forest of the Lebanon' (1 Kings 7.2), thus giving it the appearance of a fruitful garden, full of abundant life. The seven-branched candlestick, the menorah, has also been likened to the tree of life, and here too is the immense 'bronze sea' (1 Kings 7.23–26; cf. vv. 38–39) that we mentioned above. It is not an aspect of the temple to which we pay much attention, but it must have been quite imposing and a very important aspect of its symbolism, judging from its description:

[23]Then he made the cast sea; it was round, ten cubits from brim to brim, and five cubits high. A line of thirty cubits would encircle it completely. [24]Under its brim were panels all around it, each of ten cubits, surrounding the sea; there were two rows of panels, cast when it was cast. [25]It stood on twelve oxen, three facing north, three facing west, three facing south, and three facing east; the sea was set on them. The hindquarters of each were towards the inside. [26]Its thickness was a handbreadth; its brim was made like the brim of a cup, like the flower of a lily; it held two thousand baths. (1 Kings 7.23–26)

This motif was apparently deeply rooted, for it emerges in perhaps the latest book of the Bible – Revelation. According to its vision of heaven in chapter 4, 'in front of the throne there is something like a sea of glass, like crystal' (Revelation 4.6). However, the idea that God's throne is over the waters is encountered in poetic form in Psalm 29, a composition that celebrates Yahweh as the God of storm and thunder and may be one of the earliest poems in the Bible. According to v. 10,

> ¹⁰The LORD sits enthroned over the flood;
> the LORD sits enthroned as king for ever. (Psalm 29.10)

The 'flood' here is primarily the heavenly body of water from which the rain derived, but in the psalm the praises of heaven and earth merge, and his enthronement over the waters in his earthly temple is probably also in view. Indeed, it is because the 'sea' is to be found in the temple that, in the vision of Ezekiel 47.1–12, in the time of fulfilment when the new, ideal temple is constructed, a life-giving river will flow out from under the temple door:

> ¹Then he brought me back to the entrance of the temple; there, water was flowing from below the threshold of the temple to-wards the east (for the temple faced east); and the water was flowing down from below the south end of the threshold of the temple, south of the altar. ²Then he brought me out by way of the north gate, and led me around on the outside to the outer gate that faces towards the east; and the water was coming out on the south side.
> ³Going on eastwards with a cord in his hand, the man measured one thousand cubits, and then led me through the water; and it was ankle-deep. ⁴Again he measured one thousand, and led me through the water; and it was knee-deep. Again he measured one thousand, and led me through the water; and it was up to the waist. ⁵Again he measured one thousand, and it was a river that I could not cross, for the water had risen; it was deep enough to swim in, a river that could not be crossed. ⁶He said to me, 'Mortal, have you seen this?'
> Then he led me back along the bank of the river. ⁷As I came

back, I saw on the bank of the river a great many trees on one side and on the other. ⁸He said to me, 'This water flows towards the eastern region and goes down into the Arabah; and when it enters the sea, the sea of stagnant waters, the water will become fresh. ⁹Wherever the river goes, every living creature that swarms will live, and there will be very many fish, once these waters reach there. It will become fresh; and everything will live where the river goes. ¹⁰People will stand fishing beside the sea from En-gedi to En-eglaim; it will be a place for the spreading of nets; its fish will be of a great many kinds, like the fish of the Great Sea. ¹¹But its swamps and marshes will not become fresh; they are to be left for salt. ¹²On the banks, on both sides of the river, there will grow all kinds of trees for food. Their leaves will not wither nor their fruit fail, but they will bear fresh fruit every month, because the water for them flows from the sanctuary. Their fruit will be for food, and their leaves for healing.' (Ezekiel 47.1–12)

The river goes forth to water the land, nourishing the trees on its banks and fish within its waters, and possessing such life-giving properties that it will even 'heal' (that is, desalinate) the waters of the Dead Sea (Ezekiel 47.8). Even the 'trees for food' that will grow on its banks will have miraculous properties: 'Their leaves will not wither nor their fruit fail, but they will bear fresh fruit every month, because the water for them flows from the sanctuary. Their fruit will be for food, and their leaves for healing' (v. 12). Many allusions in the prophets to the time to come, the time of fulfilment, of peace, prosperity and blessing, draw on the same idea. According to Joel, a 'fountain shall come forth from the house of the LORD' (Joel 3.18); Zechariah looks forward to a time when 'living waters shall flow out from Jerusalem' (Zechariah 14.8); and in Isaiah 33 is the hope that 'there the LORD in majesty will be for us a place of broad rivers and streams' (Isaiah 33.21). However, this symbolic language is found remarkably unchanged even in the New Testament book of Revelation:

Then the angel showed me the river of the water of life, bright as crystal, flowing from the throne of God and of the Lamb through the middle of the street of the city. On either side of the river is the tree of life with its twelve kinds of fruit, producing its fruit each month; and the leaves of the tree are for the healing of the nations. (Revelation 22.1–2)

The temple and Eden

It is not coincidental that many of the themes associated with the temple are also attributable to the garden in which God himself is described as walking, the Garden of Eden, whose very name seems to mean 'plenty'.[11] This too is an ideal, divine, beautiful, luxuriant and abundant paradisiacal space, possessing the same qualities as the temple and other representations of God's dwelling-place. Once again, the river is an important feature.

According to Genesis 2.10–14, 'a river flows out of Eden to water the garden, and there it divides and becomes four branches', which apparently then water the whole earth. Two of these are the mighty Tigris and Euphrates, and a third is the unknown Pishon. 'It is the one that flows around the whole land of Havilah, where there is gold; and the gold of that land is good; bdellium and onyx stone are there', according to Genesis 2.11–12, apparently echoing the bejewelled quality of the temple and its decorations. The last river is the diminutive Gihon, the brook in Jerusalem, that nonetheless (from this temple-centred perspective) assumes a symbolic value far in excess of its actual importance. Similarly, the relatively modest Mount Zion (on which the temple stood) can itself become viewed as the highest of the mountains, the centre of the earth (Isaiah 2.2; Micah 4.1). This too is part of the temple symbolism that also informs the picture of the river flowing out from under the temple door in Ezekiel 47. It is the holy mountain at the source of the deeps, just as was the case of the home of the high god El in Canaanite mythology. The abundant, fertile and beautiful qualities of Eden hardly need to be emphasized, but these are concepts that seem to be rooted in the idea that the temple itself was the source of the life-giving streams.

We have seen that the 'sea' had an important place in the temple and that water was understood symbolically to flow out to give life to the earth. A connection between the sea and Eden is suggested by a much less well-known reference to this 'garden of God', in Ezekiel 28. Here, in an oracle against the King of Tyre, Eden is clearly presented as a specifically divine space. The wealthy maritime trading city of Tyre was situated on an island just off the Mediterranean coast to the north-west of ancient Israel, close to other Phoenician centres such as Byblos and Sidon. In this oracle the king is depicted as an Adam-figure who is cast out of Eden, but intriguingly it seems that the marine location of Tyre may have inspired the reference to the holy divine space of Eden. We shall explore the oracle a little more in Chapter 8, but for now let us focus on its references to Eden and other related terms indicating a holy place:

> ²... your heart is proud
>   and you have said, 'I am a god;
> I sit in the seat of the gods,
>   in the heart of the seas,'
> yet you are but a mortal, and no god,
>   though you compare your mind
>   with the mind of a god ...

> ¹³You were in Eden, the garden of God;
>   every precious stone was your covering,
> carnelian, chrysolite, and moonstone,
>   beryl, onyx, and jasper,
> sapphire, turquoise, and emerald;
>   and worked in gold were your settings
>   and your engravings.
> On the day that you were created
>   they were prepared.
> ¹⁴With an anointed cherub as guardian I placed you;
>   you were on the holy mountain of God;
>   you walked among the stones of fire.
> ¹⁵You were blameless in your ways
>   from the day that you were created,
>   until iniquity was found in you.
> ¹⁶In the abundance of your trade
>   you were filled with violence, and you sinned;

so I cast you as a profane thing from the mountain of God,
and the guardian cherub drove you out
from among the stones of fire. (Ezekiel 28.2–16)

Eden is here equated with 'the garden of God' (v. 13) and 'the
holy mountain of God' (v. 14; cf. v. 16): it is also the seat, or
dwelling-place, of God (v. 2), the sacred space of his abode, so
this explains the impression that Eden too may be understood as
having the qualities that are elsewhere associated with the temple.
However, it is clearly indicated in v. 2 that 'the seat of the gods [or
'God']' is 'in the heart of the seas'. It is fertile, rich and beautiful,
but the sea is also an implicit feature.

In fact, the idea of the sea as a sacred space, and in particular as
the preserve of God alone, probably finds its fullest, and certainly
its richest, expression here in Ezekiel's condemnation of Tyre.
This city was the hub of an extensive trading network, encompassing Phoenician colonies, including Carthage in North Africa
(which would eventually become more important than Tyre itself)
and ports as far away as Cadiz (possibly the biblical Tarshish),
on the Atlantic coast of Spain. The Greek historian Herodotus,
writing in the fifth century BC, even reports a Phoenician attempt
to circumnavigate Africa.[12] This has a ring of authenticity about
it, since he mentions that the sailors who made this journey had
claimed that, 'as they sailed on a westerly course round the southern end of Libya [that is, Africa], they had the sun on their right
– to the northward of them'. This is an assertion that Herodotus
himself regarded as incredible, but that we now know to be a perfectly accurate description of what happens when you cross the
equator. The immediately preceding section of Ezekiel, chapter
27, graphically illustrates the extraordinary extent of the mercantile connections of Tyre and of the opulence of the products for
which it traded. 1 Kings likewise tells us in respect of a much
earlier period that Solomon was dependent on Phoenician expertise for his attempts to run a merchant fleet on the Red Sea (in 1
Kings 9.26–28, 10.22). Since the book of Kings makes some very
exalted claims for Solomon himself,[13] this indicates the extent
to which the Phoenicians were acknowledged as supreme in this
sphere.

To understand the force of the oracle in Ezekiel 28 it is worth looking back at the context of the other Eden story, found in Genesis 2—3. It falls within the narrative block comprising Genesis 1—11, which is known as the 'primeval history'. Many of the stories of Genesis 1—11, those placed before the time of Abraham, are concerned with human sin and its punishment, and Genesis 2—3 is probably the most famous example of this. The sin of the builders of the Tower of Babel/Babylon (Genesis 11) was their hubris, their aspiration to God-like status in presuming they could build to heaven and become like gods themselves. At least part of the reason for the expulsion of Adam and Eve from the Garden in Genesis 3 was because eating of the tree of the knowledge of good and evil made them 'like God [or 'like one of us'], knowing good and evil' (Genesis 3.5, 22), and because, if they remained, there was a risk that they would eat of the tree of life and 'live for ever', thereby infringing again on a preserve that should be solely divine (Genesis 3.22).

However, more pertinent to the judgement against Tyre in Ezekiel 28 is another special quality of the divine realm, which the king of Tyre is perceived as infringing. This is beauty and perfection, as represented not just by the abundant growth and life in the temple and in Eden, but by the jewels and gold found in each, either on the priestly robes[14] and decorating the temple,[15] or within the Garden.[16] Such was the power and wealth of Tyre as it navigated the known world cultivating mercantile connections and amassing great luxuries, that – as Ezekiel saw it – it too was possessed of the same precious objects, and was 'perfect in beauty', aspiring to divine status.[17] This city's confident activities in the 'heart of the seas' were felt to be a further affront to God's divine exalted status (28.2): it was his place alone to be in the centre of reality, enthroned above the waters 'in the seat of the gods'. We can conclude from this that the heart of the seas is to be viewed as a sacred divine space, a place that is the seat of God, the location of the garden of God, the holy mountain. It is not to be trespassed upon proudly by aspirational human beings who seek wealth and glory for themselves.[18] It is a holy place, like the heavens, which are similarly outside the limits set for humanity, as the builders of the tower of Babel discover in Genesis 11.

Key message

Both the strange realities of the deep ocean that we are begin-ning to encounter through modern technology, and the biblical theology that invites an understanding of the deep seas and of the deeps beneath our feet as a sacred space stemming from God and known only to him, point in a similar direction. They invite humility and wonder at what we do not know and may never fully understand, the places we cannot enter, and respect for the boundaries between human and divine. At the same time, the centrality of the ocean in the water cycle, supporting life throughout the earth, encourages an appreciation of this water as genuinely life-giving, embodying like nothing else the blessings of God.

Challenge

Climate fixing?

One of the greatest concerns about the earth's changing climate is that we might be altering the way the water cycle works across the globe. In different parts of the world the spring of water bubbling up from the ground, the oasis in the desert or the rainy season are all important to the animals and plants, and not just the humans, who depend on these various sources of water for both food and drink. With human impacts on the climate it is likely that some parts of the globe will become drier, experiencing more severe droughts, while other parts will get wetter, with increased flooding and consequent effects. Life in some places on earth will become more precarious as a result.

In response to this, some people are suggesting that we 'fix' the planet by using geo-engineering.[19] That is, it is proposed that we should use a variety of techniques to try to cool the earth. One technique involves using large ships suitably equipped to spray seawater high into the atmosphere to provide cloud condensation nuclei. These are small particles of sea salt around which droplets of water can form in the atmosphere with clouds being the result.

Clouds reflect sunlight, and therefore more clouds would mean more reflection and so the earth would be cooled. This and some other geo-engineering schemes suffer from the problem that we cannot be sure that they would be effective. If humanity continues its use of fossil fuels, thereby adding more carbon dioxide to the atmosphere, and these schemes then fail, it could result in catastrophic rapid greenhouse warming (like the air conditioning in a house failing on a very hot day and the temperatures rising rapidly as a result).

The challenge is learning how to respect the sacred space that is the sea and the earth more generally. In Psalm 24 it says:

The earth is the LORD's and all that is in it,
    the world, and those who live in it;
for he has founded it on the seas,
    and established it on the rivers. (Psalm 24.1–2)

It is God's sacred space and not ours to do with as we will. We need to consider carefully how we respect the sacred nature of the earth.[20] This is particularly the case when we seek to embark on technological 'fixes' such as geo-engineering for problems caused by our misuse of the earth's resources.

Fast water, slow water, and virtual water

Both biblical and scientific understandings of the sea and other bodies of water, such as rivers and aquifers, encourage us to recognize each element not as distinct but as part of a single dynamic system. The water cycle, however understood, is one that is in constant motion and collectively embodies the totality of water on the planet. We saw above that from a biblical perspective the 'deep' does not just pertain to the sea, but is also understood as a rich reserve of water nourishing freshwater rivers and streams. It is easy to dismiss this as a piece of outmoded ancient geography that has no place in our modern understanding of the world. However, it is important to realize that what we talk about as 'the water cycle' is, properly speaking, only a fast water cycle taking place at the surface level of the earth. Far more water is tied up in a slow water cycle, which entails most of the earth's water being locked up in long-term reserves.

The 'fast' water cycle – with water evaporating from the seas and land, forming clouds, and dropping to earth as rain again – actually entails a very small proportion of the earth's water, and indeed most of this 'fast' water rises from the sea and falls back down over the sea as well, so is of little practical consequence for our use. By contrast, water percolating through aquifers may spend anything from a few months to a million years underground, with 1,000 years being a reasonable average, and the timescales involved with water being locked up in oceans and ice-caps are similar, so we might think of this as a 'slow' water cycle. Of course, there can be some interchange between the two, with rainwater going on to percolate through aquifers, or meltwater running out of glaciers. However, we can see that the biblical concept of water reserves – vast quantities caught up in the deeps, as well as God's storehouses of rain and ice – actually reflect in many ways the reality that most water is indeed 'stored' more than it is in rapid surface motion. The association of the 'deeps' not only with the sea, but also with nourishing fresh water, and the connection of both with the temple, suggests that consideration of the implications of our freshwater use would be appropriate here.

The statistics indicating the distribution of the world's water are very striking.[21] The earth contains about 1.4 billion cubic kilometres of water – the sort of quantity that we cannot begin to conceptualize.[22] However, since about 97 per cent is seawater, this leaves only 3 per cent (or 35 million km³) as fresh water. Two-thirds of this is contained in glaciers and ice-caps, so just one-third (about 12 million km³) is in liquid form. However, the vast majority (over 98 per cent) of this liquid water is found in aquifers, leaving only about a sixtieth of the liquid fresh water above ground, whether in lakes and rivers, permafrost and soils, marshland and swamps, atmospheric water vapour, or living organisms such as ourselves. Obviously, just a proportion of this is accessible, and water in rivers and lakes comprises only about 92,000 km³ at any one time, though of course it is not always the same water, as it is constantly on the move. About 40,000 km³ of water actually falls over land as rain in a single year and, of this, approximately 14,000 km³ is in principle capturable.[23] The further complication is that a tenth of rivers flow into the Arctic, and those with the greatest flow – the Amazon, Congo, and Orinoco – run through uninhabitable jungle for most of their length. This means only about 9,000 km³ is in practice usable. With a world population of 7.125 billion (7,125,000,000), this leaves about 1,260 km³ (or 1,260,000 litres) per person.

This sounds like a lot until you realize that it takes 2,000–5,000 litres of water to grow just 1 kilogram of rice, and 20,000 litres for 1 kilogram of coffee. Livestock production is worse: it takes 24,000 litres of water to grow enough feed to make 1 kilogram of beef, or 2,000–4,000 litres for a litre of milk, and about 5,000 litres for 1 kilogram of cheese. Indirect use can easily exceed our 'quota', even before household water consumption is taken into account.

A really critical area, then, in which we ought to consider our impact concerns the use of 'virtual water'. This is not about our personal direct use of water sourced from within the UK, but the impact of our consumer choices on less water-rich countries – for example, when we buy rice, coffee or cotton products. These crops, as we have seen, absorb vast quantities of water in the place of origin, even though they are often grown in arid, water-poor regions. They are then consumed here, although we are water-rich (if somewhat lacking in the sun they need to grow). We are in effect, then, buying 'virtual' water, indirectly consuming significant quantities of this scarce resource in places where it is needed most.

What is most alarming about this is that in most of the regions where such crops are grown, whether India or the American West, farmland is nourished through the serious depletion of river water, to the extent that so much is withdrawn for agriculture and production that sometimes previously mighty rivers (such as the Colorado) are scarcely able to trickle to the sea, or may not make it to delta regions at all. Such large-scale diversions of water are likely to deprive others of the water they need and to cause serious damage to river ecosystems and wetlands. In addition, there can be heavy reliance on the use of water from aquifers, even though these can take hundreds, thousands, or even millions of years to replenish and are therefore being depleted at a rate that will lead to exhaustion and a crisis situation for those depending on them in arid regions. If this happens, few crops at all – let alone the water-hungry ones currently cultivated at greater profit than those that might be grown just to feed the local market – will be able to survive. Even within England, where rain is a characteristic feature and a default topic of conversation, a quarter of rivers are at risk of drying out, according to a current campaign by the WWF (World Wide Fund for Nature).[24]

Discussion and reflection

- Are there limits to how much we should try to influence global ecological, chemical and physical systems by attempting to find scientific solutions to the problems we have created? How much can we know and understand, and how much do we need to know in order to begin to seek to alter it? What are the risks and benefits?
- The Bible offers a very strong sense of the sea as the preserve of God and a source of life (and death). It is something that we as humans cannot aspire to master. How can we recover a sense of the sacredness of the sea, and indeed of other elements of creation?
- It is worth spending time reflecting on water as a precious life-giving gift from God.[25] Perhaps this poem by Wendell Berry might be a way to do so?

Water
I was born in a drouth [drought] year. That summer
my mother waited in the house, enclosed
in the sun and the dry ceaseless wind,
for the men to come back in the evenings,
bringing water from a distant spring.
Veins of leaves ran dry, roots shrank.
And all my life I have dreaded the return
of that year, sure that it still is
somewhere, like a dead enemy's soul.

Fear of dust in my mouth is always with me,
and I am the faithful husband of the rain,
I love the water of wells and springs
and the taste of roofs in the water of cisterns.
I am a dry man whose thirst is praise
of clouds, and whose mind is something of a cup.
My sweetness is to wake in the night
after days of dry heat, hearing the rain.

Action

In light of the sea and the earth being a sacred space, we need to live within the place and means gifted to us, and not overstep this. We considered above technical schemes to avoid the consequences of our negative impact on the planet. However, the framework of living within limitations may also be applied to our attitude to water extraction. In the Old Testament, the drying of the deep was understood as one of the most extreme consequences of sin, as the earth mourned and suffered at human transgression of the divinely ordained order (see, for example, Isaiah 19.5–10, 24.3–13, and Jeremiah 23.10). We are now in danger of drying 'the deep' in the form of aquifers as a deliberate short-term measure to increase immediate profitability, without thought for the very serious long-term impacts. There could be no more profound illustration of overstepping the natural limits of the earth, with potentially devastating consequences.

As individuals not directly involved in climate-altering schemes or water-extraction, there are still mitigating steps we can take. First, we need to accept our own responsibility for our personal part in climate change and its limitation through changes in our lifestyles and expectations, and second, we need to think very carefully about our use of water through indirect means, as well as more obvious direct uses. This is particularly the case if we are to conceive of fresh water as a gift from God, flowing out to bestow blessing on the earth, rather than as a commodity to be exploited and (in some locations, such as where there is high dependence on aquifers) exhausted.

Changing the climate?

Climate change is not something invented by scientists. Climate change is something caused by us collectively as a species, both in the past but also in the present by our everyday collective actions. Technological efforts of climate change mitigation are only examples of further action by human beings that may radically affect the welfare of the planet. At least, however, these are intentional efforts to improve the current or projected situation for the intended benefit of all species, however risky or questionable these procedures may be.

The best solution to the problem we have created is a radical collective change of lifestyle, but societal changes begin at the individual and local level. Moreover, reducing our impact on the climate does not just concern our immediate direct actions (like overheating our homes or indulging in long-haul flights). It is also a feature of many of the small decisions we make each day, like eating beef or drinking cows' milk instead of choosing alternatives, since cattle produce significant levels of methane. In fact, livestock production, according to a UN report published in 2006, accounts for 18 per cent of global emissions,[26] and a more recent study suggests that this may be up to a quarter.[27] This means that if we cut out or reduce the amount of beef or lamb in our diet, we can make a considerable difference.[28] Similarly, we need to consider not just our own journeys, but those of the products we might buy: if we wish to consume New Zealand lamb, Chinese electronics or Californian strawberries, the climate impact of this is considerable. Lower environmental standards in certain countries in which manufacturing is consequently cheaper is another instance in which each of us as a consumer, or potential consumer, can make a difference.

Counting the drops

The other area – which we have touched on above – that requires immediate personal action is reflection on our use of 'virtual' water. A polluted trickle is a travesty of the biblical ideal of the life-giving river flowing out from the temple, gifting fertility as it flows. Therefore, the abuse and over-exploitation of water, especially where it is at the expense of the economically disadvantaged, has to be addressed. Europe is currently a net importer of virtual water, despite the fact that we do not (in general) suffer serious water shortages. By contrast, Ethiopia effectively exports thousands of gallons in coffee; in Russia, the Aral Sea has been destroyed by the draining of its water for cotton plantations in Central Asia; Israeli oranges are delightfully juicy partly at the expense of impoverished Palestinians, who are among the most water-deprived populations in the world; a third of the flow of the River Indus is used to make cotton, meaning that it no longer reaches the Arabian Sea; and there are many other examples of similar situations worldwide. We need seriously to consider the issues of justice entailed in these unequal transactions, and the impact of our consuming choices on the long-term viability of these regions, and on local people. Counting the water involved in production, and not just the carbon emissions, and – even more importantly – allowing this to influence our consumer choices, is a very important place to start.

Notes

1 The dimensions and capacity of the 'bronze sea' are the subject of a number of scholarly articles, for example, Kjell Hognesius, 1994, 'The capacity of the molten sea in 2 Chronicles IV 5: a suggestion', *Vetus Testamentum*, 44/3, pp. 349–58, and Oded Lipschitz, Ido Koch and Shlomo Guil, 2010, 'The enigma of the biblical bath and the system of liquid volume measurement during the first temple period', *Ugarit Forschungen*, 42, pp. 453–78.

2 The Hawaiian Islands were created as a result of a hotspot in the earth's mantle underlying the tectonic plate.

3 Pangaea and Panthalassa come from the Greek words *pan* meaning 'all', *gaia* meaning 'earth' or 'land', and *thalassa* meaning 'ocean' or 'sea'.

4 See Nick Lane, 2016, *The Vital Question: Why Is Life the Way It Is?*, London: Profile Books.

5 This is a biblical phrase used to describe the sources of the sea, which feed it from beneath. See Genesis 7.11, 8.2 (both in relation to the Flood) and Proverbs 8.28.

6 See Exodus 24.15—27.19, 39.1—40.33, Numbers 8.1-4, and 1 Chronicles 28.10-20. Note that the site of the Jerusalem temple was where the angel of the Lord had appeared, as described in 1 Chronicles 20.15-27 and 2 Chronicles 3.1.

7 Richard Clifford, 1972, *The Cosmic Mountain in Canaan and in the Old Testament*, Cambridge, MA: Harvard University Press, p. 114.

8 For example, at Larsa, Sippar and Babylon. See Eric Burrows, 1935, 'Some cosmological patterns in Babylonian religion', in S. H. Hooke (ed.), *The Labyrinth: Further Studies in the Relation Between Myth and Ritual in the Ancient World*, London: SPCK, pp. 43-70.

9 See John M. Lundquist, 1984, 'The common temple ideology of the Ancient Near East', in Truman G. Madsen, *The Temple in Antiquity: Ancient Records and Modern Perspectives*, Provo, UT: Religious Studies Center, Brigham Young University, pp. 53-76 (specifically pp. 59-60).

10 The name 'El' simply means 'God', and so can mean both 'a/the god' in a generic sense and can refer to the particular God of the Bible. It appears regularly in the Old Testament, over 200 occurrences of which are in respect of God himself (for example, in Numbers 12.13, 16.22, Job 5.8, Psalm 17.6 and Psalm 19.1), besides in place names such as Bethel (literally 'house of god/El') or El-Elohe-Israel ('El/God, God of Israel', Genesis 33.20), and in personal names, such as those of the father and son Elkanah and Samuel, and of course Immanuel, 'God [is] with us'. However, it is often not possible to discern when it means 'God' – the particular God of the Old Testament – or 'god' in a more general sense. The name Elijah means 'my G/god is Yah[weh]', so is a good example of this. It may be intended to assert personal loyalty to one God, Yahweh, in a context in which others may have chosen to worship other deities, such as Baal.

11 Lawrence E. Stager, 1999, 'Jerusalem and the Garden of Eden', in Frank Moore Cross Festschrift, *Eretz-Israel*, 26, pp. 183-94 (specifically p. 186). Most of the biblical references to Eden concern the idea of lush growth: see Isaiah 51.3, Ezekiel 31.9, 16, 18; 36.35 and Joel 2.3.

12 Herodotus, *Histories*, 4.42.

13 See, for example, 1 Kings 10.6-7, 20-21, 23-25, 27. Of course, Solomon is also described as being dependent on Phoenician craftsmen, and indeed on their cedar wood which they floated down the coast from the Lebanon, for the building of the Jerusalem temple (see 1 Kings 5.1-18 and 7.13-47).

14 For the jewels, see Exodus 25.7, 28.9-14, 17-21, 39.6-7, 10-14; and for gold, Exodus 28.5-6, 8, 13-14, 15, 22-27, 33-34, 36, 39.2-5, 8, 13, 15-20, 25-26 and 30.

15 For example, gold and onyx, besides all sorts of precious stones, in 1 Chronicles 29.2. For the lavish gold decoration of Solomon's temple and its furnishings, see 1 Kings 6.20-22, 28, 30, 32, 35, 7.48-51.

16 Genesis 2.11-12; Ezekiel 28.13.

17 Ezekiel 28.4–6.

18 A similar thought is expressed in Horace's *Odes*, I, 3 (published in 23 BC):

Useless for a wise god to part
the lands, with a far-severing Ocean,
if impious ships, in spite of him,
travel the depths he wished inviolable.
(Translation from www.poetryintranslation.com)

19 Alternatively, and more emotively, known as 'planet hacking' for obvious reasons. For more on geo-engineering, consult the Royal Society report *Geoengineering the Climate*, available at https://royalsociety.org/topics-policy/publications/2009/geoengineering-climate/.

20 There is no thought of seeing the earth as divine (pantheism) here, merely an acknowledgement that God is immanent in creation (as well as transcendent beyond creation).

21 We have drawn these, and the figures below relating to the water-needs of different crops, from Fred Pearce, 2006, *When the Rivers Run Dry: What Happens when our Water Runs Out?*, London: Eden Project Books.

22 A cubic metre equates to 1,000 litres. A cubic kilometre (or 1,000,000,000 m³) is enough to fill over 300,000 Olympic swimming pools.

23 One way of imagining this is to think of all the water in the world as contained in a gallon jar. Less than one teaspoon would be the usable fresh water. The following National Geographic video uses this analogy and explains the global water crisis well: www.youtube.com/watch?v=Fvkzjt3b-dU.

24 *'Water for Wildlife: Tackling Drought and Unsustainable Extraction'* (WWF, June 2017).

25 Try using this link: www.youtube.com/watch?v=34atkNBDvZY.

26 This is documented in the Food and Agriculture Organization of the United Nations (FAO) report, 2006, *Livestock's Long Shadow: Environmental Issues and Options*, Rome, available at www.fao.org/docrep/010/a0701e/a0701e00.HTM. A simple representation (from an American context) of the environmental impact of different diets can be found at http://shrinkthatfootprint.com/food-carbon-footprint-diet.

27 Sonja J. Vermeulen, Bruce M. Campbell, and John S. I. Ingram, 2012, 'Climate change and food systems', *Annual Review of Environment and Resources*, 37, pp. 195–222.

28 For a short clip indicating the different amounts of water used in the production of common foods and simple dietary choices that can make a difference, see www.youtube.com/watch?v=GOLf2RbxmzE.

6

Coping with Chaos and Uncertainty: The 'Chaotic' Sea

Chaos reigns?

Appropriately enough for a chapter on 'chaos', even the meaning of the term is confusing since it has several different senses, and we shall touch on at least three of them here. One of the most important is the scientific definition, which we shall come to in a later section, but for the time being we shall begin with two more familiar aspects of 'chaos'. One definition is that it refers to 'formless primordial matter', as the *Oxford Dictionary* puts it, and this provides a link immediately to the first chapter of Genesis. The opening lines of the Bible tell us that in the beginning 'the earth was a formless void' (Genesis 1.2; the Hebrew is *tohu wabohu* – an evocative phrase in itself) and the rest of the chapter describes God bringing order and pattern out of the apparent chaos.

However, the meaning of chaos that is of much greater contemporary concern is the idea of confusion and disorder, the discomfiting disintegration of what we know. The rest of the Old Testament (and indeed the New Testament too) is much more interested in this kind of chaos, and often conceptualizes the fear and disorientation associated with political or social instability as a roaring ocean threatening to ingress the land. There are references to waters roaring and foaming (Psalm 46.3), to the pounding waves and the thunder of the great waters (Psalm 93.3–4), to the tossing and roaring of the waves (Jeremiah 5.22), and to a violent storm (Mark 4.35–41). All of these paint a picture of a threatening and dangerous environment – that of the sea. This is an environment that is beyond human control (Psalm 107.24–30;

Mark 4.35–41), an environment that at times can seem chaotic and fraught with uncertainty. In the biblical mindset, these roaring waves also connect with the first kind of chaos, since if the waves really did cover the earth again then we would have a return to a state that was more characteristic of pre-creation, devoid of any order and structure. It is a very graphic and extreme way of expressing the fearful possibility of the loss of all we know and depend upon.

All these examples point to the common human experience that the world is neither a safe place nor an entirely predictable one, even if God is in control. From our perspective we need to learn to cope with the inevitable chaos and uncertainty that afflicts our lives and the feelings of vulnerability that this engenders.

Human vulnerability

Nothing in creation more effectively exemplifies human vulnerability before the raw power and unpredictability of natural forces than the sea. However, the fearful situation of being caught in a storm at sea also provides a metaphor for all that we struggle with, whether we are 'drowning' in a sea of paperwork, 'buffeted' by the storms and troubles of this life, or 'engulfed' and 'sinking' in the face of other woes. Even when we can maintain a serene façade, we may talk of paddling madly under the surface in order to keep going.

The Victorian poet and hymn-writer William Whiting appreciated the power of this metaphor more than most, and indeed dedicated his collection *Rural Thoughts and Scenes* as follows:

In memory
Of a dear friend,
Who, through the waves of sin and sorrow
Which flowed around him
In this ocean-world of life,
In patience and faith guided his earthly bark,
And full of charity to all,
At length sought rest in the haven where he would be

Here the dangerous 'waves' threatening his friend's course through life represent the contrary forces that, in the end, led him to seek a safe 'haven' out of their reach. However, in his most famous hymn, adopted by the British armed forces and the navies of the USA and many Commonwealth countries, Whiting did not only develop the graphic aura of threat presented by the 'restless wave'. He also introduced a pervasive sense that, despite the 'peril' of 'the mighty ocean deep', God remains fully in control, able to impose limits and to calm and protect his people 'where-so-e'er they go':

> Eternal Father, strong to save,
> Whose arm does bind the restless wave,
> Who bids the mighty ocean deep
> Its own appointed limits keep;
> O hear us when we cry to Thee
> For those in peril on the sea.

> O Saviour, whose almighty word
> The winds and waves submissive heard,
> Who walked upon the foaming deep,
> And calm amid the rage did sleep;
> O hear us when we cry to Thee
> For those in peril on the sea.

> O Holy Spirit, who did brood
> Upon the waters dark and rude,
> And bid their angry tumult cease,
> And give for wild confusion peace;
> O hear us when we cry to Thee
> For those in peril on the sea.

> O Trinity of love and pow'r,
> Your children shield in danger's hour;
> From rock and tempest, fire, and foe,
> Protect them where-so-e'er they go;
> Thus, evermore shall rise to Thee
> Glad hymns of praise from land and sea.

This hymn recognizes that turbulent forces are truly fearsome and dangerous, yet also affirms that they are ultimately kept within limits and that salvation and life are nonetheless possible. This resonates both with the physics of the natural world and with the biblical passages on which Whiting drew. However, it also offers a message of trust in God and affirmation of an ultimate level of order despite the immediate chaos. This contrasts strongly with modern secular responses, classically of panic and of intense mitigating action intended to eliminate threats to our established order. However, to understand better what it is to live with chaos, we shall need first to examine scientific understandings of 'chaos' and order, before moving on to the biblical material and then back to our own context.

The 'butterfly effect'

One might reasonably ask where butterflies feature in the oceanic world, and in a direct sense they clearly do not! However, there is a very famous scientific paper by the meteorologist Edward Lorenz entitled 'Does the flap of a butterfly's wings in Brazil set off a tornado in Texas?'[1] For the ocean, Lorenz's question about the so-called 'butterfly effect' might be re-stated as 'Does the flap of a butterfly's wings in Brazil set off a hurricane in the Atlantic, leading to devastation when the hurricane makes landfall at the coast?'

Put simply, the butterfly effect is an arresting way of expressing the fact that physical systems such as the ocean, the weather and the climate can be very sensitive to very small and unpredictable factors. Longer-term, larger-scale effects, such as the formation of a hurricane, are sensitively dependent on the initial conditions and so really can be influenced by small changes such as the flap of a butterfly's wing. Scientists refer to this phenomenon as 'chaos'. 'Chaos' in this sense is not complete randomness. Rather, it describes how a complex system, such as the weather, is so sensitive to small changes in conditions that its behaviour is so unpredictable as to *appear* random. Every element within it obeys the laws of physics, but the system as a whole is so complex and sensitive to minute variable factors that the outcome cannot be predicted. This is why the weather is usually predictable over a few days but not over periods longer than ten to fifteen days. In practice, our ability to predict the consequences of changes in natural systems will always be limited and this will vary from system to system and affect the length of time over which predictions can be made. [2]

Chaos, in the sense used by scientists, is ubiquitous. Current scientific understanding suggests that it is present not only in physical systems, but in chemical, biological and ecological ones too. It can also be present in human-designed systems such as electronic circuits. What this means is that the world that we live in is at some level unpredictable and will always be like that.[3] This does not mean that we cannot predict anything. Clearly, we design and build aircraft using computer simulations and predictions and they do not drop out of the sky.[4] However, the natural world is much more complex and therefore inherently less predictable.

The sea retreated

There is an apocryphal story of the attempt by King Canute (more correctly Cnut), the eleventh-century Viking king, to command the tide. In most versions of the story, Canute is misrepresented as being deluded and thinking he could command the sea. Actually, he did this to show his flattering courtiers that he was human and only God can command the sea, for God alone is King.[5]

In reality, when the sea dramatically retreats this is a warning of a particularly unpredictable and potentially devastating form of natural disaster: a tsunami. This is exactly what happened in the Boxing Day tsunami of 2004 in the Indian Ocean. This tsunami led to the deaths of over 230,000 people in coastal communities bordering the Indian Ocean. One young British girl, Tilly Smith, saved the lives of about a hundred tourists because she had been taught in school that a sign of the imminent arrival of a tsunami was the sea retreating from the shore. When she saw the sea receding dramatically, she alerted her parents to the danger, and they persuaded fellow holidaymakers to move quickly to higher ground before the tsunami wave struck.[6]

A tsunami is usually caused by undersea seismic activity, such as an earthquake, but could be caused by an underwater nuclear explosion or the impact of a large meteorite on the ocean.[7] Meteorite impact is a favourite of filmmakers, as in the films *Armageddon* or *Deep Impact*. The latter had the tag line 'Oceans rise, cities fall'. Tsunamis seem to be inherently unpredictable, in that some major episodes of undersea seismic activity (which are detected using seismometers) do not generate tsunamis, while other, smaller, submarine earthquakes may. Earthquakes themselves are likewise highly unpredictable, compounding the problem. Finally, in addition, how a tsunami travels across the ocean and interacts with a particular shoreline remains an active area of research and is the subject of further uncertainty.

The waters rose

Flooding is something that we in the UK have become much more aware of in recent years as it seems to have become a more common experience. Most of this flooding is due to heavy rain and rivers overflowing their banks. In contrast, those countries subject to hurricanes or typhoons or tropical cyclones[8] making landfall on their coasts may experience flooding for different reasons. In particular, strong atmospheric low pressure systems, like hurricanes, generate storm surges that can inundate coastal areas when the storm makes landfall.

The UK, the Netherlands and Belgium experienced such an event in 1953, when the combination of high tide and storm (not a hurricane) in the North Sea generated a storm surge, which caused severe flooding and deaths in the coastal areas of these countries. This led to the improvement of coastal defences in the Netherlands and the building of the Thames Barrier in the UK (although that took some time to build, only becoming operational in 1982).

However, this 1953 event pales into insignificance when compared to the devastation caused by Hurricane Katrina hitting New Orleans in the USA in 2005. The New Orleans surge protection levées failed in multiple places, leading to the flooding of

80 per cent of the city. It was the costliest natural disaster in the history of the USA and there were a number of deaths associated with Katrina. It also led to a sharpening of the divide between rich and poor, as those with transport were able to evacuate, while those without had to stay and face the storm. This underlines a general point that in cases of natural disasters it is the poor who suffer most. Storm surges such as the one that hit New Orleans are a common occurrence in the Bay of Bengal due to tropical cyclones. These badly affect Bangladesh where many poor people live in the low-lying areas near the coast.[9] Likewise, the effect of Super Typhoon Haiyan on the Philippines in 2013 was disastrous, with under-privileged people suffering the most.

Although considerable effort is expended on hurricane research, their exact path and strength are still a challenge to predict accurately. Additionally, our human defences against their effects often seem puny and inadequate.

The rains came and went

Water is critical to life of all kinds – humans, animals and plants – on earth. The evaporation of water from the ocean and rainfall over land are part of the global water cycle,[10] and this too can be unpredictable. One of the most dynamic parts of that cycle occurs in the tropical Pacific where the behaviour of the atmosphere and ocean feed back on each other and can lead to the well-known El Niño phenomenon and its counterpart La Niña.

In normal conditions in the tropical Pacific, easterly winds[11] along the equator 'push' warm water towards the western (Indonesian) end of this ocean. Evaporation from the warm water leads to rainfall occurring over Indonesia. This is an ocean-atmosphere feedback system, in that the warm water heats the air in the west which then rises and returns east at high levels in the atmosphere. The air cools and descends at the eastern end of the Pacific to feed the easterly winds. Sometimes the easterly winds weaken, for reasons that are not entirely clear, and the warm water starts to move east across the tropical Pacific towards the Americas. This affects the feedback loop so the easterly winds weaken further and the warm water moves further east, eventually reaching the coast of the Americas. As the warm water moves away from the Indonesian end of the tropical Pacific, the rainfall moves with it. This can lead to devastating forest fires in Indonesia, as the rains are no longer there to suppress either naturally occurring fires or those resulting from human activity.

The effects of El Niño are widespread in terms of changes in rainfall patterns. Droughts in Australia and Africa, as well as heavy rainfall and flooding on the west coast of the USA, are some of the most noticeable impacts. Another consequence is an increase in tropical cyclones due to warmer waters in the central Pacific providing more energy for their formation.[12] All these changes affect human beings, sometimes severely, particularly as the rains come and go.

La Niña is the counterpart to El Niño, where the easterly winds strengthen (again for poorly understood reasons) and 'push' more warm water to the Indonesian end of the tropical Pacific and even beyond into the Indian Ocean. Its effects on human populations are generally less severe.

The name El Niño means 'the Christ Child' – a name given by Peruvian fishermen in the 1800s to a warm current appearing at the coast near Christmas. The fishermen noticed this because it led to a drop in the abundance of fish. The warm waters overlying the colder waters that usually upwell at the coast cause this drop in fish numbers. The colder waters bring nutrients from deeper in the ocean, promoting the growth of plankton on which the fish feed. The effects of El Niño have in the past devastated the Peruvian fisheries with disastrous consequences for the people who rely on them for their livelihood.

The behaviour of El Niño is not well understood despite many years of research. It is known to occur irregularly every two to seven years, but we still lack a clear grasp of the causes. For example, based on satellite observations of sea surface temperatures, a large El Niño was thought to be developing in 2014, but in fact it fizzled out. In contrast, subsequently in 2015–16, a large El Niño event took place in the tropical Pacific with the effects being felt in places like Papua New Guinea, where there was a lack of rain. Such changes are beyond human control but can have devastating impacts on people living in the areas affected by either too little or too much rain. They may also be beyond our ability to predict accurately, due the butterfly effect.

An uncertain world – on the edge of chaos now

What we can see from the examples of tsunamis, hurricanes and El Niño is that the world is an uncertain place and that any illusion we have of being in control is just that – an illusion. The forces at work in creation are far beyond human control and often defy our attempts to protect ourselves from them. This should give us pause to think about our humanity and vulnerability in this uncertain world.

However, though the world is uncertain and many things in it are inherently chaotic and unpredictable, it remains an ordered world nevertheless. The laws of physics apply. For example, try jumping off a tall building and flying and the law of gravity will soon disabuse you of the notion that you can! The world seems to be balanced on the edge of uncertainty. Too regular a world would probably mean that life could not exist. Too chaotic a world would probably mean the same. For example, when we breathe in, the air enters our mouth from all directions but when we breathe out it leaves in a narrower turbulent flow. Otherwise, we would continually breathe the same air in and out (assuming we did not move around) and soon exhaust its oxygen content. Similarly, genetic variation and mutation has allowed the process of evolution, which is necessary for life on earth to adapt to different environments and for new species to emerge. Without it we would not be here, and there would have been no development from the simplest life forms. However, if living cells were too prone to mutation, genetic disease and cancers would be much more prevalent, so the evolution of life exists on the edge of chaos too. It is worth noting that the same physics that allows tsunamis, hurricanes and other destructive phenomena to occur is also responsible for the less dramatic benign effects that benefit us. These include the weather patterns that bring the rain, hence allowing crops to grow, and the ocean currents that provide nutrients for phytoplankton growth, thereby creating an abundance of fish, and so on.

So life might be said to exist on the edge of chaos. The interplay between uncertainty and regularity is what makes the world an interesting place and indeed also an inhabitable one. We know that winds generate ocean waves that will eventually break on some distant beach. The surfer paddling out on a surfboard to 'catch' a breaking wave does not know which one will give a good 'ride' back on to the beach, but the anticipation and the experience of trying to surf such a wave is exhilarating and fun. However, that same interplay of regularity and uncertainty can also be dangerous for us. Even in the case of surfing, there might be an undercurrent that, due to water running off the beach, might drag the surfer under and drown her. We live in an uncertain world.

An uncertain world – on the edge of chaos back then

The people of the Bible were much more acutely aware of the possibility of chaos than we tend to be now. Although much of what we have said so far about tsunamis and ocean currents was beyond their scientific understanding – and indeed, largely outside their geographical awareness – they lived in a culture in which the vast majority of people eked out a subsistence existence. As a result, they were particularly vulnerable to variations in natural cycles: droughts, floods, earthquakes, locust plagues, and disease suffered either by the people themselves or by their crops or livestock. Each of these 'chaotic' events could lead to grave suffering or even death.

On the political level too the Bible reflects a constant struggle for survival as an independent state. It presents a picture of the group of small tribes during the period of the Judges constantly being in tension and warfare with neighbouring peoples – Mesopotamians, Canaanites, Amorites, Ammonites, Moabites, Amalekites, Midianites and Philistines (not to mention the lesser-known Sidonians, Perizzites, Hivites and Maonites, among others) – while the monarchy itself seems to have arisen in the face of the Philistine threat. Over the following millennium, Syria, Assyria,

Babylonia, Persia, Greece and Rome each successively asserted their power over the region, with regular interference from Egypt at times exercising a pincer effect on the Israelites and Judeans caught in between.

However, in biblical times 'chaos' was understood not just as experienced in these aspects of life in the Levant, but as a fundamental aspect of the world. The earth itself was understood as founded on water and surrounded by water, above, below and all around. This was essential for life: the rains nourished the earth, and the waters beneath did the same by welling up as rivers and streams. At the same time, the sea that surrounded the earth was also understood as presenting a constant threat. This is well illustrated by the brief Psalm 93:

> [1]The Lord is king, he is robed in majesty;
>> the Lord is robed, he is girded with strength.
> [2]He has established the world; it shall never be moved;
>> your throne is established from of old;
>> you are from everlasting.
> [3]The floods have lifted up, O Lord,
>> the floods have lifted up their voice;
>> the floods lift up their roaring.
> [4]More majestic than the thunders of mighty waters,
>> more majestic than the waves of the sea,
>> majestic on high is the Lord!
> [5]Your decrees are very sure;
>> holiness befits your house,
>> O Lord, forevermore. (Psalm 93.1–5)

What is very striking about this psalm is the way it both acknowledges the very real threat of the 'floods' (more literally, 'rivers') and 'mighty waters' that lift up their voice and roar, and at the same time it affirms Yahweh's much greater power and kingship. There are real dangers in the world and yet it is 'established' and 'shall never be moved'. God's laws and holiness too are guarantors of an orderly and meaningful cosmos.

God of all

In fact, the Bible affirms that God not only stills and confines the sea, but also stirs it up and makes it roar. He is Lord of all the earth, and all that happens is under his control, a vast canvas on which he plays out his purposes. In the following passage from Jeremiah 31.35–36, God's work to 'stir up the sea so that its waves roar' is seen as part of his action in upholding the 'fixed order' of creation, akin to providing light through the sun by day and through the moon and stars at night. Here, then, the roaring sea is a sign of good order in the world, not of threatening chaos. This order is so firm that it is offered as a guarantee of God's work in ensuring the continuation of his nation Israel:

> 35Thus says the LORD,
>   who gives the sun for light by day
>   and the fixed order of the moon and the stars for light
>   by night,
>   who stirs up the sea so that its waves roar –
>   the LORD of hosts is his name:*
> 36If this fixed order were ever to cease
>   from my presence, says the LORD,
>   then also the offspring of Israel would cease
>   to be a nation before me forever. (Jeremiah 31.35–36)

[*These last two cola recur in Isaiah 51.15, again in a context where the message is not to fear but to trust in God's commitment to his people and in his promise of salvation.]

If we turn to Psalm 107, we see that God can both cause sea storms and still them:

> 23Some went down to the sea in ships,
>   doing business on the mighty waters;
> 24they saw the deeds of the LORD,
>   his wondrous works in the deep.
> 25For he commanded and raised the stormy wind,
>   which lifted up the waves of the sea.

²⁶They mounted up to heaven, they went down to the depths;
  their courage melted away in their calamity;
²⁷they reeled and staggered like drunkards,
  and were at their wits' end.
²⁸Then they cried to the LORD in their trouble,
  and he brought them out from their distress;
²⁹he made the storm be still,
  and the waves of the sea were hushed. (Psalm 107.23–29)

In Psalm 89, it is again affirmed that God rules the sea, though here this is seen primarily in his work in stilling its rising waves:

You rule the raging of the sea;
  when its waves rise, you still them. (Psalm 89.9)

A similar idea occurs in Jeremiah 5:

Do you not fear me? says the LORD;
  Do you not tremble before me?
I placed the sand as a boundary for the sea,
  a perpetual barrier that it cannot pass;
though the waves toss, they cannot prevail,
  though they roar, they cannot pass over it. (Jeremiah 5.22)

This very thought is picked up thousands of years later in the hymn with which we began:

Eternal Father, strong to save,
Whose arm does bind the restless wave,
Who bids the mighty ocean deep
Its own appointed limits keep

In both cases, God's power to restrain and limit the sea provides assurance of his wider powers to maintain order and stability in the world or indeed in people's lives. In all these passages, there is an implicit message that Yahweh is King. He is not only in control of natural processes but can also be trusted to preserve his people. It is meaningful to say 'O hear us when we cry to Thee / For those in peril on the sea' because he listens and responds.

The roaring nations

In a further set of Old Testament passages, the terrifying nations that so often invaded Israel and Judah are likened to the sea transgressing the land. Psalm 65 affirms:

> You silence the roaring of the seas,
>   the roaring of their waves,
>   the tumult of the peoples. (Psalm 65.7)

Here the threatening sea merges into the equally intimidating armies of the 'peoples'. We might need to be reminded that loud or constant noise – the machinery, alarms and traffic that form a backdrop to the lives of so many of us today – had no part in the ancient world and was associated almost solely with thunder and warfare. However, the analogy of enemy peoples with the sea also had a comforting aspect: God as Lord of creation stills and sets bounds for the sea. This vast and most dynamic and fearsome aspect of the natural world is governed and restrained by him. How much more therefore can he limit the action of enemy peoples.

This idea is expressed especially clearly in Isaiah 17.12–13, where the thought of the nations being rebuked and fleeing away like waters receding before God turns into the realization that they are like chaff in the wind before him:

> ¹²Ah, the thunder of many peoples,
>   they thunder like the thundering of the sea!
> Ah, the roar of nations,
>   they roar like the roaring of mighty waters!
> ¹³The nations roar like the roaring of many waters,
>   but he will rebuke them, and they will flee far away,
> chased like chaff on the mountains before the wind
>   and whirling dust before the storm. (Isaiah 17.12–13)

What is most striking is how these ancient writers found meaning and comfort as they grappled with their greatest fears not by underplaying the danger of what faced them, but by affirming God's control.

Here be dragons

We have already looked at a number of passages where fear of chaos is expressed in terms of the world order disintegrating as water engulfs the land. This is clearly given a political emphasis in the many examples where an analogy is understood between threatening waters and overwhelming enemy forces. However, in possibly even more extreme cases, the object of fear is thought of as embodied in a 'dragon'. It is not directly claimed that God battles dragons now, but he is imagined to have had a victory over such a beast (or beasts) in the past, and this provides hope for the present. It is also anticipated that he will triumph decisively over this terrible beast in the future.

The writer of Psalm 74, faced with the devastating crisis of the destruction of the temple, recalled how in the glorious past,

> [13]You divided the sea by your might;
>> you broke the heads of the dragons in the waters.
> [14]You crushed the heads of Leviathan;
>> you gave him as food for the creatures of the wilderness.
>> (Psalm 74.13–14)

Isaiah 51.9, which comes out of the same crisis of exile to Babylon and the destruction of Jerusalem, likewise pleads with God, referring to his slaying of a dragon named Rahab, rather than Leviathan:

> Awake, awake, put on strength,
>> O arm of the LORD!
> Awake, as in days of old,
>> the generations of long ago!
> Was it not you who cut Rahab in pieces,
>> who pierced the dragon? (Isaiah 51.9)

In each instance, the overcoming of the dragon was a decisive act in the past through which God redeemed his people, and now he is petitioned to re-enact a similarly decisive victory in order to save them once again. Probably both references allude to his triumph over the Egyptians at the Red Sea, when God parted the waters and the Israelites crossed safely but the sea overwhelmed their enemies. Indeed 'Rahab' at times clearly refers to Egypt (as in Isaiah 30.7 and Psalm 87.4) which suggests it may well be in mind here too in Isaiah 51.[13] However, this is not *just* a simple mention of Egypt, because here that dominant nation is conceptualized as a terrible beast that was slain.

In these passages, God is invoked to enact a similar victory in the present, liberating his people from bondage in a foreign land. He had once slain the 'dragon' Egypt to release his people from slavery and now they need a similar miracle in order to be freed from oppression by the Babylonian Empire, which had taken them into captivity once more. However, in a much later crisis, when salvation through normal historical processes no longer seemed conceivable, the idea of God's decisive victory over Leviathan still offered hope in a desperate situation. The writer of Isaiah 27 affirms:

On that day the LORD with his cruel and great and strong sword will punish Leviathan the fleeing serpent, Leviathan the twisting serpent, and he will kill the dragon that is in the sea. (Isaiah 27.1)

Somehow on that day, the decisive day in the future when Yahweh will intervene, the dragon will be overcome and right order will be established again. This dragon was doubtless understood as an oppressive enemy power even if the battle is painted in colours that rather transcend the here-and-now realities of human conflict or even of typical divine action against hostile nations.

This imagery was to be extended further in the book of Revelation. Here the dragon becomes a more comprehensive embodiment of evil, one not only found especially in the tyranny of Rome, but also identified with the Devil and Satan, who is the one who deceives the nations:

[1]Then I saw an angel coming down from heaven, holding in his hand the key to the bottomless pit and a great chain. [2]He seized the dragon, that ancient serpent, who is the Devil and Satan,

and bound him for a thousand years, ³and threw him into the pit, and locked and sealed it over him, so that he would deceive the nations no more, until the thousand years were ended. After that he must be let out for a little while ...

⁷When the thousand years are ended, Satan will be released from his prison ... ¹⁰And the devil ... was thrown into the lake of fire and sulphur, where the beast and the false prophet were, and they will be tormented day and night forever and ever. (Revelation 20.1–10)

Here the object of fear is not diminished but, as with the earlier images of roaring water and the dragon, acknowledged as a terrifying and very real threat. This is not something that might be tackled through human effort, but through reliance on God alone.

In quietness and trust

When faced with a crisis, our natural instincts are often to rush into action to protect ourselves. This was seen in the move to bomb Syria following the Paris attacks in November 2015, and arguably the Iraq war was at least in part motivated by a desire to do something (or at least to feel and look as if something was being done) in response to 9/11. US president Donald Trump's promise on his election to exclude Muslims from the USA plays directly into those same fears and a desire to take radical action to protect oneself from perceived threats. Sometimes such action is appropriate and well considered, and at other times less so. However, action of itself offers us a sense of control, whether we are slaughtering livestock (and even the occasional pet pig) in order to stem the spread of foot and mouth, or pumping funds into finding a vaccine for the latest threatened human pandemic (AIDS, avian flu, Ebola, Zika ...), or heightening security and giving police forces stronger powers to detain suspected terrorists, or even directing scientific research into attempts to find technological solutions that will allow us to maintain our current levels of dependence on carbon-based energy without simultaneously fuelling global warming. We can affirm order for ourselves, we can work to maintain and reinstate the status quo, and we can announce above all that the forces of chaos will not restrict our freedom.

Of course, action was understood as providing the solution to overcoming chaos in an ancient context too, but such activity was primarily understood as being enacted in the divine sphere. Ultimately, from a biblical perspective, the guarantor of stability is God. The Bible affirms that the earth may be shaken but it cannot be moved because he is King and Creator. He established it firmly so that it will endure for ever. The floods may roar and lift up their voice but God is mightier than them and will still them. The nations may come in the clamour of war, but safety is only in him. Evil may appear to have taken over the world, and yet God will ultimately intervene and destroy the dragon and Satan.

The reality, of course, was that people then, as now, felt the need to protect themselves and to immerse themselves in frantic activity. This tension is expressed especially clearly in Isaiah 30:

> [15]For thus said the Lord GOD, the Holy One of Israel:
> In returning and rest you shall be saved;
> > in quietness and in trust shall be your strength.
> But you refused
> [16]and said,
> 'No! We will flee upon horses' –
> therefore you shall flee!
> and, 'We will ride upon swift steeds' –
> therefore your pursuers shall be swift! (Isaiah 30.15–16)

Isaiah urges the Judeans to rely on God, not on military might. They think they should depend on their own resources, exemplified by the use of horses (an expensive and prestigious asset in ancient warfare), but Isaiah counters that this would only lead to defeat. In another passage (8.6–8), he urges trust in the quiet, life-giving waters of the Temple Mount (representing trust in God himself), not in alliances with the overwhelming floodwaters of the dominant superpower of the period, Assyria.

This particular message of judgement in Isaiah 8.6–8 is addressed to the Judean king, Ahaz. It comes from the time when the king of Damascus, Rezin, and the king of Israel, Pekah son of Remaliah, tried to force Judah to join them in an anti-Assyrian coalition during a period when the region was heavily under threat from this expanding empire. The Judean king panicked and called on Assyria for help, and of course Assyria was only too happy to oblige. This is the message that Isaiah delivered into this context:

> [6]Because this people has refused the waters of Shiloah [in Jerusalem] that flow gently, and melt in fear before Rezin and the son of Remaliah; [7]therefore, the Lord is bringing up against it the mighty flood waters of the River [Euphrates], the king of Assyria and all his glory; it will rise above all its channels and overflow all its banks; [8]it will sweep on into Judah as a flood, and, pouring over, it will reach up to the neck; and its outspread wings will fill the breadth of your land, O Immanuel. (Isaiah 8.6–8)

The Euphrates was the main river of Assyria, and indeed its economy and agriculture was dependent on its floodwaters. Here, though, the flooding tendencies of this great river (which is sometimes in the Bible referred to as 'The River' and sometimes as a 'sea') are used as a metaphor for Assyrian imperial expansion into the Levant. By contrast, the small channel running from the Gihon spring in Jerusalem, the 'waters of Shiloah', is the antithesis of this mighty force, yet the water flowing from the holy city also stood for the presence of God in their midst. It is in this that the people of Jerusalem and their king Ahaz should have trusted. There is heavy irony in the name used to address Judah, 'Immanuel', since it means 'God is with us'. The implication is clearly that they ought to have lived as if God really were with them, and to have trusted in his protection.

Anthropic order and anthropic chaos

The chaos we have considered so far in the biblical passages is the threat of what might be termed 'external' disorder: threats coming from outside the Judean or early Christian community, whether through floods, invasion, imperial rule, religious persecution or some other ill. Similarly, we have seen how in the physical (and indeed biological and other) systems of the earth there is unpredictability built into an overarching order, which can make life very precarious.

However, there is a further cause of chaos, recognized from both scientific and biblical perspectives: anthropic (human-caused) chaos. Our devastating effects on biodiversity, the climate and other systems is well known. However, from a biblical perspective, already there is a sense that flourishing in the natural world stems from justice and righteousness, and that human wickedness (or indeed goodness) can have ramifications far beyond this single species.

Justice ultimately comes from God, and it is this that 'keeps steady the pillars' of the earth, ensuring a stable created order.

A connection between God's justice and order in the world is made in Psalm 96:

> Say among the nations, 'The LORD is king!
>   The world is firmly established; it shall never be moved.
>   He will judge the peoples with equity.' (Psalm 96.10)

Here God's equitable judgement provides assurance of the maintenance of order in the physical world. However, the rule of a just king can have a similar effect. In Isaiah 11.1–9, it is the rule of a righteous king who possesses the spirit of God that inaugurates the glorious new era of peaceful coexistence between species. It is almost as if the king is acting on God's behalf, as his proxy:

> ¹A shoot shall come out from the stump of Jesse,*
>   and a branch shall grow out of his roots.
> ²The spirit of the LORD shall rest on him,
>   the spirit of wisdom and understanding,
>   the spirit of counsel and might,
>   the spirit of knowledge and the fear of the LORD.
> ³His delight shall be in the fear of the LORD.
>
> He shall not judge by what his eyes see,
>   or decide by what his ears hear;

⁴but with righteousness he shall judge the poor,
  and decide with equity for the meek of the earth;
 he shall strike the earth with the rod of his mouth,
  and with the breath of his lips he shall kill the wicked.
⁵Righteousness shall be the belt around his waist,
  and faithfulness the belt around his loins.

⁶The wolf shall live with the lamb,
  the leopard shall lie down with the kid,
 the calf and the lion and the fatling together,
  and a little child shall lead them.
⁷The cow and the bear shall graze,
  their young shall lie down together;
  and the lion shall eat straw like the ox.
⁸The nursing child shall play over the hole of the asp,
  and the weaned child shall put its hand on the adder's den.
⁹They will not hurt or destroy
  on all my holy mountain;
 for the earth will be full of the knowledge of the LORD
  as the waters cover the sea. (Isaiah 11.1–9)

[*Jesse was the father of David, hence 'a shoot from the stump of Jesse'
would be a descendent of David. The royal line had become a truncated
stump at the fall of Jerusalem, when the monarchy had been cut off and
Judah's existence as an independent state had come to an end, but this king
would begin a new era of royal rule.]

Here, harmony on earth stems from just rule in the human sphere.
But maintaining justice and ensuring righteousness on the earth is
not the responsibility solely of God or the king: it extends to the
whole of society.

More importantly, injustice and wrongdoing can have just as
powerful an effect as righteousness. In Isaiah 24, the results of
general human wickedness are even described in terms that imply
cosmic collapse. The windows of heaven are opened as at the
flood, so that devastating waters pour on to the earth beneath
and the whole foundations of the world shake:

¹⁸... For the windows of heaven are opened,
    and the foundations of the earth tremble.
¹⁹The earth is utterly broken,
    the earth is torn asunder,
    the earth is violently shaken.
²⁰The earth staggers like a drunkard,
    it sways like a hut;
  its transgression lies heavy upon it,
    and it falls, and will not rise again. (Isaiah 24.18–20)

Human violence and transgression are here perceived as bringing
into jeopardy cosmic order itself. Conversely, the flourishing of
all species in the land of Israel is dependent on the people's justice
and obedience. Only from this can universal blessing and order
flow. Both the negative and positive aspects of this can be illustrated from the book of Hosea. Let us take the negative example
first, from Hosea 4:

¹Hear the word of the LORD, O people of Israel;
    for the LORD has an indictment against the inhabitants of
        the land.
  There is no faithfulness or loyalty,
    and no knowledge of God in the land.
²Swearing, lying, and murder,
    and stealing and adultery break out;
    bloodshed follows bloodshed.
³Therefore the land mourns,
    and all who live in it languish;
  together with the wild animals
    and the birds of the air,
    even the fish of the sea are perishing. (Hosea 4.1–3)

Here disobedience to God's commandments affects all of creation
– the earth (or land)¹⁴ and all the creatures. By deliberately echoing Genesis 1.28, 'Be fruitful and multiply, and fill the *earth* and
subdue it; and have dominion over *the fish of the sea* and over
the birds of the air and over every *living thing* that moves upon
the earth' [italics ours], Hosea expresses how humanity's relation

with God, and hence with the rest of creation, has gone horribly wrong. (The 'wild animals' of Hosea 4.3 are more literally 'living things of the field', so are described in the same language as in Genesis 1.28, even though it is not apparent from the English translation. Similarly, the word translated 'land' in Hosea 4.1 is the same as that rendered as 'earth' in Genesis 1.)

However, the same prophet, Hosea, also has a vision of hope of a restored relationship with God, in which he gifts fruitfulness along with a renewal of his marriage with Israel:

> ¹⁹And I will take you for my wife forever; I will take you for my wife in righteousness and in justice, in steadfast love, and in mercy. ²⁰I will take you for my wife in faithfulness; and you shall know the LORD.

> ²¹On that day I will answer, says the LORD,
>   I will answer the heavens
>   and they shall answer the earth;
> ²²and the earth shall answer the grain, the wine, and the oil,
>   and they shall answer Jezreel;
> ²³and I will sow him for myself in the land ... (Hosea 2.19–23)

Righteousness and blessing may ultimately stem from God, but human reciprocity, facilitating this flow of good things on earth rather than hindering it, is a vital part of how right order on earth was experienced.

Key message

It can be seen that the delicate balance between chaos and disorder is built into creation, and indeed creation could not exist without it. The established order in the world is both dangerous and vulnerable; it cannot be taken for granted, yet through God it is ultimately sustained and overall God's rule prevails. The sense of unpredictability in the world, of danger going hand-in-hand with mechanisms that also bring wellbeing and opportunity within an overarching ordered system, is echoed by scientific understandings of 'chaos'. Another characteristic of the world, which is particularly highlighted by our understanding of the 'butterfly effect', is the incredible level of sensitivity built into the natural systems on which we depend. Systems such as the weather, climate and oceans are both extraordinarily complicated and interconnected in a way that our ancestors could not have begun to imagine, even though the Bible shows a firm grasp of this principle. The balance and welfare of all parts of creation are interconnected: one part cannot be out of kilter or change without having an impact on another. The Bible, as well as recognizing the fragility of the created order and its dependence on God, also perceives how human sin can have far-reaching effects on the 'order' of creation with negative consequences. The world is indeed remarkable but at the same time highly vulnerable and sensitive to human action.

Challenge

Given that so much of life depends on the interplay between regularity and chaos, the first challenge concerns our thinking about the world. Chaos permits life but it also creates a level of risk that is built into the cosmos. This is something that we can find difficult to accept. If we are confronted with a dynamic earth exploding into volcanoes as it renews itself, it challenges us to recognize our own vulnerability before great forces over which we have no control. The mutations that allow physical organisms gradually to adapt to their environment brought all the species on earth today into being through gradual evolution from the simplest life forms. However, the same process can also enable cancerous cells to destroy the life of someone we love – or even ourselves – and this raises all sorts of problematic questions of theodicy for us. The Bible knows nothing of the science of either process. However, its writers knew plenty about the fragility of human existence and about the needful dependence of all creation on God for its sustenance and continued stability.

This perspective suggests that living in such a world is best accomplished through trust in God's ultimate maintenance of order, despite apparent disorder at times impacting on our lives.

However, it also invites an acceptance that there is much in the world that is and will remain, quite rightly, outside our control. Nonetheless, it is worth reflecting that it is often our way of life that is incompatible with natural processes and that can exacerbate their effects. An erupting volcano is a beautiful and awesome sight, something that might give us (from a safe distance) privileged access to the deepest workings of the earth. It becomes problematic only if we seek to occupy land in a region subject to volcanic activity, and its effects can be devastating if they are exacerbated by high population densities. This is perhaps even more the case with earthquake zones, where, with appropriate construction methods, the effects of even quite significant tremors can be considerably less damaging than where housing is not designed to be resistant to movements in the earth's crust.

We have already described above the experience of being caught in a storm at sea, something that can create a huge sense of vulnerability. However, for some of the inhabitants of the ocean, this awe-inspiring phenomenon can apparently be cause for delight, as whales playfully ride the waves as if treated to an impromptu funfair. The popularity of surfing (despite its dangers) is reflective of a human wish to experience something of the power and beauty of the waves and to feel part of this magnificent aspect of creation.

Reflection and discussion

- This chapter has explored some of the paradoxes of inhabiting a world in which chaos is both necessary for life and also at times inhibitive to it. This is part of living on a dynamic earth that has its dangers as well as its delights, but that is none the less overall finely tuned for life. On a 'safe' changeless earth, life would not have evolved nor the mountains been replenished, yet if there were much more chaos built into its systems, life would not have been sustainable at all. How do you feel about this reality? Do you find it helpful in explaining how we might hold together faith in God and the reality of disasters that can seem to call this into question?

- The Bible describes a world that is in certain respects fragile and that is often under threat, and yet it also affirms that God can ultimately be trusted to ensure its continuation and stability. Does this mirror your own understanding? This perspective might provide reassurance in respect of natural disasters such as disease or earthquakes, but can it be applied to the very modern problem of anthropogenic climate change?
- Can the biblical idea of chaos help us to come to terms with disasters? The Anglican newspaper, the *Church Times*, ran a leading article immediately after the 9/11 attack which reflected on the symbolic and mythological dimensions of what had happened:[15]

Let there be no mistaking the symbolic power of these attacks on the World Trade Center and the Pentagon. The perpetrators of these attacks knew the power of symbols. The attack went to the symbolic heart of American culture – its economic control and military power.

And it is clear that the American political leadership understood that the battle at hand was a battle of mythical proportions. President Bush and other politicians stated over and over throughout the day that 'freedom' had been attacked that morning. An attack on the World Trade Center is an attack on freedom.

Of course this makes complete sense if you live in a world in which freedom is primarily defined as the liberty to make profits and the liberty to consume. An attack on the Pentagon is also an attack on freedom. Of course, countless people around the world would identify their oppression closely with the military and economic operations that were targeted on that fateful day.

The language of myth was also sharply in focus during the President's brief address to the nation on the evening of 11 September. This attack, he said, was intended to inflict chaos on the nation; but he was here to tell us that America was still in control. The President was in the White House, government services would be reopened in the morning, and, most importantly, 'America is open for business.'

Does this sound like a callous and irrelevant comment under such circumstances? Not at all. 'America is open for business' means the forces of chaos will not triumph because the forces of salvation are stronger. And salvation is found in an ever-expanding global economy.

The drama of Mr Bush's four minutes on television was worthy of the great myths of Babylon or any other empire. Tiamat, the sea goddess of chaos, might threaten the order, security and peace of the empire, but Marduk, the great god of order, will triumph. Freedom was attacked, Bush said: the military and economic order in which freedom is realized withstood the attack, and we must now 'go forward to defend freedom'.

But the President knows, his cabinet knows, the guy on the street knows, and we all know, deep down, that things will never be the same again. The American imperial mythology of invincibility, rooted in its economic and military hegemony, and historically proved by the collapse of Communism, has been shaken.

... For the first time in its history, an enemy of the nation had brought the pain and violence and bloodshed of war to American soil. No wonder people on the street said that it all seemed so unreal, so much like a movie rather than reality. The American mythology has no way of interpreting such an event. In terms of the myth, this simply could not have happened. But it did happen.

This raises all sorts of interesting questions. To what extent do we rely on 'myths', symbolic stories enabling us to see meaning in the world, in order to make sense of experience? How much, like Bush, do we need to feel (or proclaim) that we are in control in order to cope and carry on?

The writer here refers to Babylonian myth, in which the high god Marduk was understood to overcome 'chaos' in the shape of the goddess Tiamat (symbolizing salt water) which he slew in order to establish order in the earth, primarily for the benefit of Babylon. The Bible invites an understanding of God as ultimately in charge and guaranteeing some form of continued order, despite the existence of real threats. Is this a helpful (and perhaps underused) theological model for us today?

The question of economic supremacy as 'freedom' or 'oppression' is one to which we shall return in Chapter 8, but you might like to reflect on it here. To what extent might one person or one nation's 'order' and desirable status quo be another's oppression, and might that be the case in the West's economic and cultural dominance?

Action

Perhaps 'action' is the wrong term in this instance. There is little that we can do to change the way creation works and we will always as human beings experience life on the edge of chaos and live with some degree of uncertainty. However, accepting that chaos is part of reality and that our control is limited is something to which we may need to devote more thought. Knowing that we are finite and vulnerable and ultimately not in control of our destiny could have a profound effect on our sense of humility and on how we live, as well as on how we respond to inevitable disasters. Perhaps, too, it should deepen our sense of dependence on God, as providing certainty for an uncertain world.[16] This attitude of mind is beautifully expressed in the following extract from *Carmina Gadelica*, drawn from traditional Gaelic folklore,[17] though attaining such a level of trust may be much harder to live out in practice than a superficial reading of its words suggests:

Helmsman: Blest be the boat.
Crew: God the Father bless her.
Helmsman: Blest be the boat.
Crew: God the Son bless her.
Helmsman: Blest be the boat.
Crew: God the Spirit bless her.
All: God the Father
God the Son
God the Spirit
Bless the boat.

Helmsman: What can befall you
And God the Father with you?
Crew: No harm can befall us.
Helmsman: What can befall you
And God the Son with you?
Crew: No harm can befall us.
Helmsman: What can befall you
And God the Spirit with you?
Crew: No harm can befall us.
All: God the Father
God the Son
God the Spirit
With us eternally.
Helmsman: What can cause you anxiety
And the God of the elements over you?
Crew: No anxiety can be ours.
Helmsman: What can cause you anxiety
And the King of the elements over you?
Crew: No anxiety can be ours.
Helmsman: What can cause you anxiety
And the Spirit of the elements over you?
Crew: No anxiety can be ours.
All: The God of the elements,
The King of the elements,
The Spirit of the elements,
Close over us,
Ever eternally.

However, alongside developing a deepened sense of trust and dependency, we ought also to be pro-active in ensuring that natural disasters have less devastating consequences, particularly for the poor. We need to understand the world better and to order the way that we live more effectively so as to minimize the risks from natural hazards while still enjoying the beauty and majesty of God's creation. This will involve education programmes and investment in infrastructure. If there is any action that we can take in this respect it is perhaps to encourage governments to pay attention to the vulnerability of the world's poor to the effects of natural disasters[18] and to take appropriate action to protect and help them.

Finally, some of the biblical passages we have considered draw a direct connection between human behaviour and its impact on the whole ecosystem. Goodness, justice and righteousness bring blessing, but sin impacts much further than its perpetrators in bringing harm. We may not understand blessing in quite the same way and will be more diffident about identifying areas of divine judgement. Nonetheless, the environmental impact of our behaviours – and this often includes questions of justice and right living – cannot be denied.

Notes

1 The paper, the full title of which is 'Predictability: does the flap of a butterfly's wings in Brazil set off a tornado in Texas?', was presented before the American Association for the Advancement of Science in 1972, but is widely downloadable online.

2 For a popular presentation of the idea of chaos, see James Gleick, 1987, *Chaos: Making a New Science*, New York: Viking Books.

3 That is, without even going into the complex issue of quantum indeterminacy. See John Polkinghorne, 2002, *Quantum Theory: A Very Short Introduction*, Oxford: OUP, 2002.

4 Obviously aircraft *do* drop out of the sky occasionally, but this is usually due to pilot or other human error, mechanical failure, bad weather, or terrorist action. They fly because the laws of physics and aerodynamics work.

5 For God's and Jesus' command of the sea, see Chapter 1. One of us (Meric) lives in Southampton, the supposed location of King Canute's attempt to command the tide.

6 See www.telegraph.co.uk/news/uknews/1506286/Honour-for-young-girl-who-saved-tourists-from-tsunami.html or https://en.wikipedia.org/wiki/Tilly_Smith.

7 Meteorite impact is thought to be one possible cause for the extinction of the dinosaurs.

8 Tropical cyclones are referred to by the Spanish term 'hurricane' in the Atlantic and by the Chinese word 'typhoon' (literally, 'big wind') in the western Pacific. 'Cyclone' is more commonly used for such storms in the Indian Ocean.

9 There are currently about 10 million people living within a metre's elevation of sea level.

10 With regard to the global water cycle, note that 86 per cent of global water evaporation occurs from the ocean, and 78 per cent of rainfall falls over the ocean. Therefore, the ocean loses more water through evaporation than it gains through rainfall. The excess evaporated water is carried by atmospheric weather systems and provides rainfall over the land, then re-enters the ocean through rivers running into the sea.

11 That is, winds blowing from east to west.

12 Tropical cyclones form over the warm waters of the tropical ocean because they can derive energy from the evaporation of water from the ocean surface.

13 In Isaiah 30.7, the parallelistic style of poetry suggests that 'Rahab' broadly equates to Egypt:

For Egypt's help is worthless and empty,
therefore I have called her,
'Rahab who sits still.'

In the first part of Psalm 87.4, 'Rahab' is listed along with various other nations:

Among those who know me I mention Rahab and Babylon;
Philistia too, and Tyre, with Ethiopia ...

14 In Hebrew, the word 'erets, used in this verse, means both 'land' (as in the land of Israel) and earth (the world). Either or both may be implied here.

15 Brian Walsh, 2001, 'Lamenting the end of an empire', *Church Times*, 21 September 2001, p. 8.

16 This is not to imply that God is part of the world. He is clearly transcendent, but also, and perhaps from a human perspective most importantly, immanent in our world.

17 Collected and translated by Alexander Carmichael of Lismore. The first two volumes were published in 1900, and the remaining four volumes were finished by others after his death in 1912. It is debated how much Carmichael was faithful to his sources, rather than polishing them to provide a literary more than a literal rendition, but it can hardly be doubted that sea prayers were used by the inhabitants of the Western Isles.

18 Strictly speaking, there are natural events rather than natural disasters. They become disasters when they affect people negatively, quite often due to human sin (for example, in not keeping to building regulations when constructing buildings in an earthquake zone, leading to deaths when the buildings collapse). On this topic, see R. S. White, 2014, *Who Is To Blame? Disasters, Nature and Acts of God*, Oxford: Monarch.

7

The Vast, Vulnerable Sea

Psalm 104, which celebrates God as Creator and paints a vivid picture of his creation, speaks in v. 24 of 'the sea, great and wide' (alternatively translated as 'vast and spacious'). This coheres with our knowledge that the oceans cover 71 per cent of the earth and are easily seen as vast and spacious. The same aspect of the sea is captured in the first four lines of Barry Cornwall's poem 'The Sea':

> The sea! the sea! the open sea!
> The blue, the fresh, the ever free!
> Without a mark, without a bound,
> It runneth the earth's wide regions round

Both the poem and the biblical verse have that sense of the immensity of the sea, and over the centuries that has been humanity's perception of the oceans. Yet, despite its vastness, the Bible also speaks of the waters of the sea disappearing (Job 14.11) and of God drying up the sea (Nahum 1.4), which suggests its vulnerability too. That vulnerability has also become apparent in oceanographic studies, as the effects of global warming (temperature rise and ocean acidification) and pollution (especially our disposal of plastics in the sea) have been seen to influence the ocean on a global scale. The sentiment in the poem above, 'without a mark', no longer seems true. This thought of damage to the sea is captured in the last two lines of the first verse of Owen Seaman's poem 'The Uses of Ocean':

> To people who allege that we
> Incline to overrate the Sea

I answer, 'We do not;
Apart from being colour blue,
It has its uses not a few;
I cannot think what we should do
If ever "the deep did rot".'

The poet himself describes these as 'Lines written in an irresponsible holiday mood', and the tone of this composition is humorous.
However, Seaman is correct that we 'cannot think what we would
do' should the sea 'rot': more importantly, we tend not to dwell
much on this as an active possibility, and even his amusing tone
aptly reflects cultural distance from the sea, cheerful cluelessness
about its real dynamics, and blasé assumptions about its durabil-
ity, which we tend not to project on to the land in the same way.

Ironically for such a light-hearted composition, the line 'the
deep did rot' is taken from Samuel Taylor Coleridge's *Rime of
the Ancient Mariner*, which explores the serious themes of human
sin and its profound effects on the wider environment and on
the perpetrator himself, and we shall be returning to this later in
the chapter. For now, though, it is worth noting that Seaman's
'irresponsible' thoughts on 'the uses of ocean' reflect the reality
that we cannot overrate the importance of the sea to humanity
and indeed to the healthy functioning of the planet as a whole.
At the same time, his humorous tone mirrors the lack of seriousness with which we are still inclined to treat the ocean, presuming
on its vastness and convincing ourselves that anything we inflict
on it is but 'a drop in the ocean'. We have, as the poem's title
implies, made use of the sea, but now we do need to consider seriously whether we are causing the deep to 'rot'. As the poem also
recognizes, we 'cannot think what we should do' should the deep
deteriorate, and this poses an urgent challenge as to what action
should be taken. Here we explore both the vastness of the sea but
also its vulnerability.

Is the water too warm for you?

We know that the world is warming due to human use of fossil fuels (oil, gas and coal), which adds carbon dioxide (CO_2 – a greenhouse gas) to the atmosphere and heats the earth, including the oceans. As a result, part of the cause of the rising sea level is simply the thermal expansion of the water in the oceans as it warms. However, the warming of the oceans has other impacts, not least on marine life. While we might enjoy swimming in warmer waters, much ocean life is adapted to certain temperature ranges, so changes in the climate will affect it, usually adversely.

Since many marine organisms are sensitive to changes in water temperature, as the earth warms, with most of that heat (93 per cent) being absorbed by the ocean, we expect to see an effect on the marine ecosystem. Already there is evidence of warm water species of phytoplankton being found further north in the North Atlantic. As phytoplankton form the base of the oceanic food chain this will affect species such as zooplankton that feed on them, and in turn those species (such as fish) that feed on zooplankton. This will lead to a redistribution of the oceanic flora and fauna on a global scale and affect ocean ecosystems across the planet. Icelanders, for example, have already noticed changes in the location of the cod fisheries around Iceland. This might be a direct effect of changing water temperatures on the cod themselves, or an indirect effect in that the plankton that the cod feed on are now to be found elsewhere, due to the temperature changes.

One type of marine life that is very sensitive to temperature changes is coral. Corals are marine invertebrates and the best-known species are those responsible for reef building. These inhabit the tropical oceans and form hard calcium carbonate shells, with the reefs growing at about a centimetre per year. This means that many of the most beautiful reefs, such as the Great Barrier Reef, were laid down over the last two million years. Coral reefs support hugely biodiverse ecosystems, which teem with life and colour, providing a habitat for many species of fish, molluscs and crustaceans.

Tropical corals that are the basis of the reef ecosystems depend on a symbiotic relationship with algae living in their tissues. As water temperatures rise the corals respond by ejecting the algae, causing the coral to turn white. This is known as coral bleaching, for obvious reasons. The symbiotic algae that are now being ejected are what provide the corals with their vivid colours. The reasons for this expulsion are not totally clear, but the consequence is that if the algae do not return, eventually the coral will die. Human-induced warming of the oceans will impact corals and potentially cause large-scale bleaching to occur.

Does this taste too sharp?

Sometimes we add a little lemon to the water we are drinking to make it taste better, or at least less dull and 'metallic'. This makes the water more acidic but more appetizing. We are now effectively doing the same thing to the ocean on a global scale, but not with the aim of making seawater taste better. The extra carbon dioxide that is being added to the atmosphere (due to human use of fossil fuels) is, in part, being absorbed by the ocean. The dissolved gas reacts with water and forms carbonic acid, making the oceans more acidic. You may remember using litmus tests at school to determine whether a solution was alkaline or acidic, and that this was measured in terms of pH. Ocean measurements have shown that the pH of seawater is decreasing, meaning that the seas are becoming more acidic. It is as if we were conducting a chemistry experiment on a global scale.

What are the implications of this change in ocean acidity? As we saw above, many species of marine life are sensitive to temperature, but many are likewise sensitive to the acidity of the waters in which they live. We also observed earlier that some species of coral produce calcium carbonate, thereby forming reefs, and several species of ocean phytoplankton also build shells out of calcium carbonate, as do some molluscs. Unfortunately, calcium carbonate will dissolve in an acidic environment so the increasing acidity of the oceans will affect, at a minimum, marine life that makes use of this compound. Tropical corals, those amazing zones of beauty and biodiversity, are expected to be the worst affected in terms of stability and longevity. The stress caused to the corals by the increasing acidity may also lead to coral bleaching, just as increasing temperature does. In addition to these effects, the increasing ocean acidity may affect the metabolic rates and immune responses of some marine creatures. It is unclear how well the various species and ecosystems that are affected will be able to adapt or acclimatize to these changes in acidity.

The levels of ocean acidity are set to increase over the next 100 years and will exceed those found in the ocean over hundreds of thousands of years previously. Ocean acidification is not something easily reversed, even if we stop using fossil fuels and reduce carbon dioxide in the atmosphere. It will take tens of thousands of years for the ocean chemistry to adjust the acidity back to pre-industrial[1] values of pH. The ocean may be vast, but it is proving vulnerable to human influence on a global scale with unknown long-term consequences.

Is this too chewy to eat?

The sea has been used for waste disposal by human beings throughout recorded human history. When the world's population was small the impact of such behaviour was minimal or, at worst, constrained to a few localities. As the population has increased so has our impact on the oceans, and now human pollution of the sea is a major problem. Again, though vast, the sea is increasingly vulnerable to the actions of earth's dominant species – us. Sources of pollution include fertilizer washed off the land into rivers and, ultimately, the sea; direct discharge of effluent into the sea (think of the sewage pipes you sometimes see running out to sea on some beaches); ships at sea (particularly oil tankers) washing out their bilges; various oil tanker and oil rig disasters that have discharged huge amounts of oil into the ocean; and so on. The fertilizer washed off the land can cause eutrophication. This is a process whereby the resulting increase of nitrate or phosphate in the water encourages algae growth. Unfortunately, the algal species that grow rapidly under these conditions are often harmful ones that contain toxins that then enter the oceanic food chain. Such toxins can even harm or kill humans who eat the contaminated fish or shellfish.

Perhaps the defining oceanic pollutant of the modern era is plastic. Plastic is one of the most useful materials ever invented and is now ubiquitous across the world. However, the plastics that end up in the sea can cause damage to marine wildlife, with large pieces – such as plastic bags – entangling turtles, sharks or birds, or sometimes even being eaten by them, leading to their death. Much of the plastic accumulates in the so-called great garbage patches of the ocean. These lie at the centre of the great ocean gyres in the subtropical regions of the ocean (regions that lie roughly between 20° and 40° north and south of the equator).

Most of the larger pieces of plastic deteriorate in the ocean environment due to the action of waves and the sun, but they are not biodegradable in the sense of ceasing to have an impact on the environment. Rather, the majority of the plastic in the ocean ends up as small particles, microplastics, pieces of plastic perhaps only a few microns (millionths of a metre) or smaller in size.[2] Marine creatures of all kinds – zooplankton, marine worms, mussels, fish, turtles – easily ingest these very small particles. Unfortunately, microplastics are not good for marine life. The various chemicals in the plastics can cause physiological damage to the creatures themselves, but they also contaminate the food chain – with largely unknown consequences. We have little idea of the effects that microplastics in the ocean will have on life there in the long term.

The fragility of life

In the ancient world, the capacity for human beings to exercise a significant impact on the marine environment, even in a localized sense, was small. There were no synthetic materials such as fertilizers and plastics; technological limitations meant that efforts at controlling the environment had marginal effects (note that the miraculous catch of fish resulted in a haul of only 153 fish, according to John 21.11); and the population – even in urban areas – was still quite modest. This all restricted the scope for damage.

Moreover, in the wider ancient Near Eastern cultural environment, whether in Canaan or Babylon, the belief that the sea was not just a natural force, but a powerful deity at war with other divine beings, such as Baal, the storm god, was a given. The Bible itself contains numerous references to the sea as a powerful agent lifting up its voice and its waves in a way that threatened the land, so that God himself had to exercise control to limit it and protect the order of creation that he had established.[3] The authors of the Bible could therefore be forgiven for imagining that the sea did indeed have an infinite capacity to absorb anything cast into it and to stay immutable, remaining unmarked by the action of any other agents, human or otherwise.

However, there was an important aspect of the experience of life in ancient Palestine that countered these tendencies, and this was simply the fragility of human existence. This was a subsistence economy, with a climate and landscape that, in many parts, was not conducive to flourishing agriculture. Its location on the crossroads of trade routes linking Egypt with the rest of the Fertile Crescent and with Anatolia made a peaceful existence on the political level difficult to sustain. As a result, even in good years, the production of food, resistance to meteorological disaster, invasion, disease, wild animals, crop failure, drought, flood, locusts and many other natural challenges, could not be assumed. The writers of the Bible understood something that we have forgotten: that life was fragile and that the order and stability we in the modern West tend to take for granted – the plentiful availability of food and water; general good health and long life for most people; political stability and the exercise of law – was not inevitable. As a result, even though the sea was recognized as vast and dangerous, it was also understood as intrinsically vulnerable and subject to the effects of forces like the weather or the activity of great marine creatures. More concerning still, it could be impacted quite drastically by the judgement that results from human sin.

Disappearing seas

Fundamental to the biblical picture of the sea, and indeed of the created order itself, was that it cannot be taken for granted or be assumed to be immutable. A good example of this is provided by Job 14, in which the sea is thought of as mortal, just like human beings:

> [11]As waters fail from a lake,
>   and a river wastes away and dries up,
> [12]so mortals lie down and do not rise again;
>   until the heavens are no more, they will not awake
>   or be roused out of their sleep. (Job 14.11–12)

You will notice that the word 'sea' does not appear here in the NRSV, because in these verses the word that usually denotes the sea has been translated into English as 'lake'. This is faithful to the Hebrew, because it does not distinguish between what we would identify as a sea (such as the Mediterranean or Red Sea) and other big bodies of water such as an inland lake. Even a large river such as the Nile or the Euphrates could in Hebrew be thought of as 'sea'.[4]

However, in recognizing that even large bodies of water can be finite and fallible, this brief allusion reveals a worldview that understands their ultimate fragility. The writer realizes that they may be subjected to other forces like drought, but he also appreciates how lakes and other water resources cannot be relied on as wholly and unalterably secure. The image of failed water therefore does not just provide a metaphor for human mortality. It also speaks of the vulnerability of the human life that is dependent on such water reserves, and indeed of the fragility even of the sea itself.

One might add that the thought of the sea drying up is not merely a quaint reflection of an ancient worldview, but a reality in our own time. As we saw in Chapter 5, this is the case especially of the Aral Sea (between Kazakhstan and Uzbekistan), which used to be the fourth largest body of inland water in the world, but which has reduced in volume to about 10 per cent of what it was

50 years ago due to the withdrawal of water for human use. The very considerable fish population it once supported has all but disappeared and the lake itself has been reduced to a few hypersaline ponds. Although not depleted quite to the same degree, the surface area of another inland sea, the Dead Sea, has now reduced to little more than half of its 1930 level (at 600 km², down from 1,050 km²) as a result of the diversion of water from the Jordan, with a corresponding drop in the water table in the region.[5]

If we think of the Hebrew for 'sea' as encompassing large rivers, the reality of their potential for severe depletion is played out in many of the great river basins of the world that have been subject to damming and diversion in order to satisfy the needs of growing cities in areas of short water supply. The result has often been the reduction of the lower stretches of such rivers to a polluted trickle, much to the detriment of wildlife and to the ecology and economy of their wetlands and estuaries that were once teeming with life.

Naturally, there are many more allusions to sea being dried up as the result of direct divine intervention, most often in relation to the Exodus, or in anticipation of its renewal. However, Job 14.11 is significant in indicating that the drying of a sea could also comprise part of 'normal' reality independently of specific acts of intervention.

A shining wake

We have already met Leviathan, the great beast of the sea, in Chapter 3, and discovered from Job 41 how he is both fearsome and dangerous – but also an object of divine pride and delight. This amazing quasi-mythological animal is portrayed as invincible and without equal in creation. He is immune to all weaponry and, according to Yahweh, 'king over all that are proud' (perhaps to be understood as 'king over all the majestic wild beasts'). It is perhaps not surprising that he is also understood to have an impact on the sea:

[31]It makes the deep boil like a pot;
  it makes the sea like a pot of ointment.
[32]It leaves a shining wake behind it;
  one would think the deep to be white-haired.

(Job 41.31–32)

The context informs us that the foaming wake he leaves glistening behind him and the turbulence he causes in the water are signs of his immense power and size. This is something that cannot be achieved by any other creature and is indicative of his unrivalled status in creation. Nonetheless, this portrayal echoes natural phenomena and may be rooted in part in knowledge of crocodiles, whales, or other large aquatic animals whose activities may leave an immediate imprint on the water around them. It may also reflect the experience of observing luminescent marine phytoplankton,[6] which would respond to the turbulence created by the movement of a large beast by creating a wake that, at night, would indeed shine.

The imprint of Leviathan is clearly perceived as beautiful. This is obvious in the case of the shining pathway, but, according to the Bible, a white head too is the 'glory'[7] of the aged,[8] a 'crown of glory' gained in 'the way of righteousness' (that is, through a righteous life).[9] As a sign of respect, the people are required to rise before a grey head and honour the aged, and even to die 'in good old age' is expressed using this specific term for the silvery head of the elderly: in Hebrew, the righteous die 'in good grey hair'.[10] So the sea reveals the passing of Leviathan without itself being damaged, instead revealing its own honour and beauty. Nonetheless, the fundamental insight into the sea's impressionability is important: Leviathan has an effect on the sea, even if it is not a detrimental one.

The wind-tossed wave

It is not surprising that the sea is also understood as affected by the weather. For example, in Job 38.30 the deep (which was thought to feed the sea as well as freshwater sources) is described as freezing. A further natural force that is perceived as having an impact on the sea is the wind. In James 1 a person who is wavering in their faith is compared with a wave that is driven and tossed by the wind:

> ⁶But ask in faith, never doubting, for the one who doubts is like a wave of the sea, driven and tossed by the wind; ⁷, ⁸for the doubter, being double-minded and unstable in every way, must not expect to receive anything from the Lord. (James 1.6–8)

This provides a remarkable counterbalance to the common Old Testament perception of the sea as lifting up its waves in pride and roaring, so much so that it needed to be restrained by God. In wider Greek culture, too, a wave could be conceptualized metaphorically as representing forces that were impossible to resist: a sea of troubles,[11] or dreadful misfortune,[12] or strife,[13] a wave of horsemen,[14] or even a hostile wave.[15] Here in James, by contrast, a wave is perceived as weak and ephemeral, buffeted by forces greater than it, against which it has no power to resist. The waves are not a driving power and outworking of the sea itself, but objects that are themselves driven and tossed where the wind will blow. This metaphor picks up beautifully on the root meaning of the Greek word translated here as 'doubt', which is essentially to 'judge to and fro'. The wave is, then, tossed about like an undecided mind.

Waters of life, waters of death

The changes to which the sea may be subject can, however, also be positive, as is most clearly evidenced in Ezekiel 47. We met this passage in Chapter 5, where we examined Jerusalem temple theology, with its idea of the deep, the source of life-giving water, being located in (or under) the temple. Here, in Ezekiel's idealized vision for the restored temple, the waters of the river flowing out from under the door of the temple are envisaged as entering the Dead Sea and thereby making the water fresh. In the Hebrew conceptualization, however, this is not simply a question of diluting the sea's extreme salinity in order to enable it to support life. Rather, the Dead Sea – which seems to be described as the 'bitter' or 'stagnant' sea – is 'healed'. This of course makes tangible the divine blessing and healing that is embodied in the river. The result is an abundance of piscine life and the flourishing of fishing, but miraculously the swamps and marshes will not be 'healed' in order that they may be left for salt.

In Hebrew thinking, though, blessing is not simply an outworking of all that flows from the temple. It is also the result of human obedience: if the people live according to the demands of God, in justice, fairness and righteousness, then *shālôm*, wellbeing for all creation, will result. Conversely, disobedience leads to the disruption of all that *shālôm* represents: the right functioning, wellbeing and blessedness of creation. Human behaviour is thus seen as a vital element in ensuring the continuity of God's blessing in the land. If injustice, idolatry and wickedness prevail, how, when everything is so wrong, could wellbeing result?

In the modern world, we have seen all too vividly how the antithesis of the waters of life, in the form of flows of polluting effluent or leakages of gallons of oil, can bring death and destruction. Ezekiel's vision, then, not only offers the promise of healing and life to be brought even to the hyper-saline and lifeless Dead Sea. To the modern reader, it also stands as a poignant reminder not just of what could be, but of how the opposite, rivers of death, have at times – and to our shame – polluted and destroyed life in even very vibrant parts of God's creation.

The wounded sea

We have already explored how a sea may be subject to other forces: imprinted by great beasts, buffeted by the wind, dried up in its entirety, or even 'healed' when it is unable to support life of itself. These images present a valuable perspective on the sea as intrinsically impressionable, rather than solely as a force exerting itself on others.

However, there is a further set of passages in which the sea is envisaged as suffering as a result of human sin when God

intervenes in judgement. The sea is not, of course, typically singled
out for destruction. Rather, judgement can affect all of creation
and the sea is not excluded from this. In addition, in some cases,
an impact specifically on the sea is a corollary of action against a
particular nation. For example, in Ezekiel 26, the stones of Tyre
are envisaged as being cast in the sea as part of her destruction:

> They [the invading Babylonian army] will plunder your riches
>   and loot your merchandise;
> they shall break down your walls
>   and destroy your fine houses.
> Your stones and timber and soil
>   they shall cast into the water. (Ezekiel 26.12)

Given that the Tyrians were an island people, surrounded by the
sea, this image is hardly surprising.

The drying of the 'sea of Egypt' or Nile is another common
motif. This echoes the drying of the Red Sea at the Exodus, when
– according to probably the most important tradition in the Old
Testament – the Israelites escaped slavery in Egypt and fled across
the miraculously parted waters of the sea, leaving the Egyptian
forces to drown as the waves returned. However, the motif of
the drying of the 'sea of Egypt' also reflects the importance of the
Nile (which was referred to as a 'sea' in Hebrew) for Egypt's agriculture and economy:

> And the LORD will utterly destroy
>   the tongue of the sea of Egypt;
> and will wave his hand over the River [the Euphrates]
>   with his scorching wind;
> and will split it into seven channels,
>   and make a way to cross on foot. (Isaiah 11.15)

> They shall pass through the sea of distress,
>   and the waves of the sea shall be struck down,[16]
>   and all the depths of the Nile dried up.
> The pride of Assyria shall be laid low,
>   and the sceptre of Egypt shall depart. (Zechariah 10.11)

The huge economic and social impact of damage to the Nile is
elaborated in great detail in an oracle against Egypt in Isaiah 19:

> ⁵The waters of the Nile will be dried up,
> and the river will be parched and dry;
> ⁶its canals will become foul,
> and the branches of Egypt's Nile will diminish and dry up,
> reeds and rushes will rot away.
> ⁷There will be bare places by the Nile,
> on the brink of the Nile;
> and all that is sown by the Nile will dry up,
> be driven away, and be no more.
> ⁸Those who fish will mourn;
> all who cast hooks in the Nile will lament,
> and those who spread nets on the water will languish.
> ⁹The workers in flax will be in despair,
> and the carders and those at the loom will grow pale.
> ¹⁰Its weavers will be dismayed,
> and all who work for wages will be grieved.
>
> (Isaiah 19.5–10)

It must be noted that many of these images are hyperbolic and are
often not expected to be fulfilled literally. Indeed, often there is an
important metaphorical level of meaning that transcends the surface level of the text. The idea of the new Exodus, signalling the
fresh redemption of God's people and the start of a new era in his
relationship with them, underlies many of the images of the drying
of the sea. In addition, when it is said of Babylon that 'I will dry
up her sea and make her fountain dry' (Jeremiah 51.36), water is
here seen as a source of life and wellbeing – just as in Proverbs
5.15–18 a wife is also described as a fountain and as flowing
water. When this life-source is removed, desolation results:

> ³⁶Therefore thus says the LORD:
> I am going to defend your cause
> and take vengeance for you.
> I will dry up her sea
> and make her fountain dry;

37and Babylon shall become a heap of ruins,
  a den of jackals,
an object of horror and of hissing,
  without inhabitant. (Jeremiah 51.36–37)

Of course, both Babylon and Egypt were heavily dependent on the rivers Nile and Euphrates for their prosperity, and indeed for their survival. The careful management of floodwaters and irrigation of crops through man-made channels was in each case essential. The feeding of the populations of these early urban civilizations depended on such measures, and the economies and even existence of their cities were tied to effective water-management. In fact, it seems likely that the scale of work needed to construct and maintain complex agricultural irrigation not only maintained these urban civilizations, but actually brought them to birth in the first place. The motif of the drying of the 'seas' of the Euphrates and Nile (as in Isaiah 11.15, 19.5, quoted above) strikes at the foundations of the great powers of Babylon and Egypt, but also plays into the concept of nourishing, irrigating water being a blessing from God.

The idea that the sea could be dried up – and indeed had been dried up decisively at the Red Sea – or that it could otherwise be severely damaged is a highly significant one. It reflects human dependence on water and the realisation that it is a gift and a source of divine blessing, rather than an inexhaustible 'given' that can be taken for granted or viewed as a right. It also reveals an appreciation that all creation is interdependent and that where there is imbalance and suffering in one element, it may have a deleterious effect elsewhere – and indeed, possibly everywhere. Human sin therefore inevitably impacts on the land and sometimes also on the sea.

'Water, water everywhere, and all the boards did shrink'

Perhaps the best-known picture of the drying of the sea in the English language comes not from the Bible, but from a poem profoundly influenced by Christian thought. The above quotation, drawn from Samuel Taylor Coleridge's poem *The Rime of the Ancient Mariner*, offers a much-quoted picture of becalmed sailors surrounded by water on the ocean, but without a drop to drink:

> Water, water, every where,
> And all the boards did shrink;
> Water, water, every where,
> Nor any drop to drink.

However, perhaps less well known are the following two stanzas in which it is made clear that the sea itself is envisaged as a drying swamp:

> The very deep did rot: O Christ!
> That ever this should be!
> Yea, slimy things did crawl with legs
> Upon the slimy sea.
>
> About, about, in reel and rout
> The death-fires danced at night;
> The water, like a witch's oils,
> Burnt green, and blue and white.

Underlying this imagery is a deep sense of the interconnectedness of humanity with the rest of the created world, and the moral framework informing our relations with other creatures. In order to understand this better, however, we need to look at the broader context of this particular 'lyrical ballad'.

Within the poem, the Mariner is depicted as telling his compelling story to a wedding guest who is transfixed by the tale 'like a three-years' child', hence the reader is implicitly invited through this narrative device to take the guest's part and suspend disbelief, entering into the sailor's fantastical mindset in order to absorb the Romantic truths of the poem's message without being required to examine its supernatural premises. The poem is too long to quote here (though it is easily found online), but, to summarize it briefly, the 'Mariner' tells the story of how he had embarked on a long voyage during which he was blown south to the Antarctic. Blessing seemed to accompany the ship when it was followed by an albatross whom the sailors befriended, whereas becalming and the death of all but the Mariner himself ensued after he shot the bird for no apparent reason. Of crucial importance, though, are the Christ-like aspect of the albatross, whom the mariners hail as a 'Christian soul', and the irrationality and terribleness of the deed of shooting this bird.

'God save thee, ancient Mariner!
From the fiends, that plague thee thus! –
Why look'st thou so?' – With my cross-bow
I shot the Albatross.

This deed has profound spiritual and material consequences, and seems to take on the character of the Fall, just as redemption comes through the cross. However, a crucial aspect of the poem is also that the shooting of the albatross grieves

The spirit who bideth by himself
In the land of mist and snow,
He loved the bird that loved the man
Who shot him with his bow.

Within this poem, natural bodies such as the sun, moon and storm are constantly personified, creating an impression of (super-) natural powers determining the course of events. This polar spirit too seems to be the force that determines what ensues, guided by love despite the terrible lessons that must be learned in order for the deed to be atoned. As we have seen, after the shooting of the albatross, it is not merely the ship that is becalmed and the sailors who are parched for lack of water, since even the sea itself is profoundly affected.

Redemption for the Mariner finally comes through spontaneous feelings of love for creation. The sailors had seen the albatross in functionalist terms, as bringing them good or ill luck, but for the Mariner release comes in spontaneous blessing of his fellow creatures:

Beyond the shadow of the ship,
I watched the water-snakes:
They moved in tracks of shining white,
And when they reared, the elfish light
Fell off in hoary flakes.

Within the shadow of the ship
I watched their rich attire:
Blue, glossy green, and velvet black,
They coiled and swam; and every track
Was a flash of golden fire.

O happy living things! no tongue
Their beauty might declare:
A spring of love gushed from my heart,
And I blessed them unaware:
Sure my kind saint took pity on me,
And I blessed them unaware.

He finds beauty in creatures that a few lines earlier were objects of horror. For the Mariner, guilt acts as a barrier to prayer, and love for creation enables release. There follows further penance, expiation and confession, though he is condemned to the 'living death' of guilt and the memory of the horror he and his crewmates had endured, as well as with the painful compulsion to tell his tale to those who need to hear it: 'the man hath penance done,/ and penance more will do'. The power of the poem lies partly in its ability to communicate in highly imaginative terms the horror of sin and suffering, of isolation and guilt, and the importance of expiation and penance. There is no easy shortcut to healing, and the terrors the Mariner had experienced will be with him always and will profoundly change all whom he meets. The wedding guest who had initially wished to escape to join the celebrations was first transfixed by the tale and then left with no appetite to join his fellow guests.

He went like one that hath been stunned,
And is of sense forlorn:

A sadder and a wiser man,
He rose the morrow morn.

However, the poem also makes a profound contribution to the idea that a deed trivially done can have immense consequences both for the sinner and for a far wider web of life, and to the recognition of the crucial value of non-human life and the essential character of love for all living things. Its central message is that:

He prayeth best, who loveth best
All things both great and small;
For the dear God who loveth us,
He made and loveth all.

It offers a distinctively mystical appreciation of the spiritual level of interconnection between humanity and other creatures, and between apparently isolated deeds and consequences that draw in the (super)natural forces of sea, sun and moon, among others. As such, it provides an intriguing bridge between modern, rationalist modes of responding to a world in which we see the sea being impacted by human action, and biblical images of sin and consequences that also seem to operate on a super-human scale. Above all, it invites us to take sin – and especially sin against creation – seriously and to be chastened by how its burden cannot easily be shaken off.

Understanding divine action

As English speakers, we draw a sharp distinction between action and consequence, sin and punishment. However, the Hebrew language views these processes much more organically. The word for 'sin' can of course denote the wrongdoing itself, but it can also encompass the consequences of that sin in the guilt[17] or punishment[18] that results. It is all part of a single process that cannot be viewed as a series of independent elements that in principle might be detached from one another. This is illustrated well by Hosea's metaphor of sowing the wind and reaping the whirlwind (Hosea 8.7).[19] Of course, sowing and reaping are distinct activities, but sowing renders reaping inevitable: it has started a process of which this is the only likely end. Sowing seeds also determines exactly what shall be reaped: the two are organically linked. You cannot sow thistles and expect to reap wheat. This process does not require special intervention, but is an inbuilt consequence of something initiated by the sowers themselves.

The idea of judgement therefore must not be understood simply as a special intervention of God from outside an otherwise balanced and functioning world in order to impose punishment. Rather, it is something that begins from within and impacts outwards on to other parts of creation. God in the Bible is understood as constantly acting in the world in a very immediate way, personally managing the tides and seasons, the circuits of the sun and moon, the provision of rain and growth of crops. This is not sharply to be distinguished from his action in judgement and blessing, which must not be understood simply as something imposed from outside, but as in many ways operating organically within the world order that God had established. Judgement and blessing are consistent and just and are initiated by the human actions, whether in obedience or sin, that set in train certain consequences. Judgement and blessing are enacted by God personally, of course, but the initial cause is also human.

An innocent victim?

Nonetheless, these passages of judgement, in which the sea is dried and its fish killed, do seem worrying. The language may be metaphorical, the suffering described may never have happened, and the problem may have begun with human sin, but still these references are challenging. The primary difficulty is with the apparently instrumental use of the sea. It seems to suffer the death of its fish as well as being made like a desert, not because of its own wrongdoing but because of the sin of the Egyptians or Babylonians, and specifically in order to make these powerful peoples suffer.

Probably such passages should be understood in terms of the

land mourning or flourishing as a result of the behaviour of its inhabitants. Thus, if the Egyptians contravene God's will, the land of Egypt itself will exhibit the symptoms of this malaise. Sadly, too, it is a reality that the consequences of harmful behaviour do not always, or even often, inflict the most damage on the perpetrators. This is true of a huge plethora of ills, from climate change to, say, drink driving or bullying and abusive behaviour, so it is possible to envisage the environment as likewise suffering the consequences of the sin of the people. More importantly, there seems to be a strong sense of identity between a people and its land in the Bible.

There is a deep sense of logic, then, in the land, or watercourses, suffering along with the sinners living there, and this is reflected in many of the oracles of blessing and judgement addressed to Israel itself. The land mourns and suffers as a result of the sins of its inhabitants, and flourishes when wellbeing results from 'welldoing'. To put it in other terms, if Egypt is punished for sin, it is not unnatural for the 'Sea of Egypt' to suffer as a result because it *is* Egypt – or at least a vital part of it. The problem of instrumentality – the sea suffering as a result of another's ills – only arises if we create a dichotomy between land and people. If they are closely identified, then this in many ways evaporates. In fact, in many of these passages, this point could be taken a stage further. The Nile or Euphrates are often metonyms. That is, they themselves represent and embody the nations of Egypt and Babylon: they stand for something wider than the rivers themselves, just as when we speak of 'the crown' we rarely have in mind a distinctive form of headwear worn by the Queen.

Further challenges

Of course if we step back into the logical world of the twenty-first century there is a problem with this thinking, much as it may have been perfectly coherent within the context of, say, the sixth century BC. Most of us would want to object somewhat to the identification between people and land, simply because we are not the only species on this earth. Why should innocent fish or other life forms suffer as a result of human behaviour? Of course, this is one of the hardest questions that we need to confront in modern times, in the wake of the current pressing environmental issues.

The paradox is that the suffering of other creatures, and even of huge natural bodies like some of our largest rivers or lakes, as a result of human behaviour is a reality today: it is not just a product of outmoded ancient thinking, but an ever more pressing problem. The fact is that we do have a far greater impact on other species, and indeed on the whole planetary system, than any other creature that has ever lived, and we need to recognize that and respond to it, rather than simply protesting on the sidelines that it should not be so. We need to live in the image of God in order to enable all of creation to flourish as it should.

An apocalyptic scenario

Of course, the worst pictures of damage to the sea occur in Revelation, where heaven and earth, sea and dry land – and indeed all creation – are envisaged as subject to destruction. Here we shall look at two brief passages where the greatest devastation seems to be envisaged, in Revelation 8.8 and 16.3. Revelation is possibly the hardest biblical book for us to get to grips with, and not only because of its disturbing content. Its highly symbolic imagery is culturally far removed from our modern Western preferences, which are in danger of being more narrowly literalistic than the writings of any other age. In Revelation, what is foreseen transcends the boundary between reality and unreality: it portrays events that are anticipated, but also quite unlike anything ever previously experienced. Despite its compelling graphic imagery, much of what it envisions is intended as symbolic, with concealed levels of meaning that are difficult for us to access.[20]

The first of our passages needs to be read in context, since it is one of a series of devastating events heralded by a trumpet blast and enacted by a destroying angel:

⁶Now the seven angels who had the seven trumpets made ready to blow them.

⁷The first angel blew his trumpet, and there came hail and fire, mixed with blood, and they were hurled to the earth; and a third of the earth was burned up, and a third of the trees were burned up, and all green grass was burned up.

⁸The second angel blew his trumpet, and something like a great mountain, burning with fire, was thrown into the sea. ⁹A third of the sea became blood, a third of the living creatures in the sea died, and a third of the ships were destroyed.

¹⁰The third angel blew his trumpet, and a great star fell from heaven, blazing like a torch, and it fell on a third of the rivers and on the springs of water. ¹¹The name of the star is Wormwood. A third of the waters became wormwood, and many died from the water, because it was made bitter.

¹²The fourth angel blew his trumpet, and a third of the sun was struck, and a third of the moon, and a third of the stars, so that a third of their light was darkened; a third of the day was kept from shining, and likewise the night. (Revelation 8.6–12)

The second passage describes a similar sequence of destruction, though this time more complete:

¹Then I heard a loud voice from the temple telling the seven angels, 'Go and pour out on the earth the seven bowls of the wrath of God.'

²So the first angel went and poured his bowl on the earth, and a foul and painful sore came on those who had the mark of the beast and who worshipped its image.

³The second angel poured his bowl into the sea, and it became like the blood of a corpse, and every living thing in the sea died.

⁴The third angel poured his bowl into the rivers and the springs of water, and they became blood ...

⁸The fourth angel poured his bowl on the sun, and it was allowed to scorch people with fire; ... (Revelation 16.1–4, 8)

The imagery of Revelation 8.8, in which 'something like a great mountain, burning with fire, was thrown into the sea', epitomizes cataclysmic cosmic disturbance as a result of divine judgement and pervasive human evil in the world. The same is true of the bowl of wrath in Revelation 16.3, and in each case the whole cosmos is affected: the earth first, then, second, the sea, followed by the rivers and freshwater sources, and next the sun, moon and stars. Nothing escapes, especially when it is made clear that death ensues in each of these areas: the plants and the sea creatures, as much as the land and sea themselves, are devastated.

For the sea to turn to blood implies profound cosmic disturbance, and (as in the case of the plagues of Egypt in which water was also turned to blood) pollution and death. The signalling of each episode in Revelation 8 with the blast of a trumpet indicates a process of assault, since in the ancient world trumpets were used to call troops into the attack. At each blast there is a fresh onslaught against successive parts of creation. This is cosmic warfare. In 16.3, the devastation is even more total, since 'every living thing in the sea died'.

Another important aspect of Revelation 8.8, though, is that the mountain itself is destroyed as well as causing destruction. The mountain may actually represent 'Babylon', a cipher for Rome, the epitome of the evil, destroying kingdom that wreaks devastation on the earth. In Jeremiah 51 Babylon is depicted as a mountain that will be destroyed in judgement:

> ²⁴I will repay Babylon and all the inhabitants of Chaldea [another name for Babylon] before your very eyes for all the wrong that they have done in Zion, says the LORD.
> ²⁵I am against you, O destroying mountain, says the LORD,
> that destroys the whole earth;
> I will stretch out my hand against you,
> and roll you down from the crags,
> and make you a burned-out mountain.
> ²⁶No stone shall be taken from you for a corner
> and no stone for a foundation,
> but you shall be a perpetual waste,
> says the LORD. (Jeremiah 51.24–26)

There is good reason to suspect that the imagery employed here in Revelation 8.8 picks up on that in Jeremiah 51, since in Revelation 17.9–10 seven mountains are identified with seven kings. Similarly, Revelation 16.20 states that 'every island fled away, and no mountains were to be found', and this again seems to be bound up with destruction of Babylon/Rome and the nations (v. 19). The image in Revelation 8.8 therefore powerfully illustrates both the destructive and self-destructive nature of evil. The action of Babylon/Rome, as it 'destroys the whole earth' (16.25), polluting and devastating the sea, has an especial character of self-destruction since Rome depended on the sea for the continuation of the commercial activities that sustained and prospered the empire. In Revelation 16, too, the specific judgement against 'those who have the mark of the beast' (v. 2), the people representing the forces of evil, could not be more clear – and yet destruction goes much further than this, as the death of everything in the sea reveals.

What can we draw from these passages? First, despite Revelation's ultimate message of hope, and of new heavens and a new earth, there is no limit to the suffering that can result from human sin. This is something to which no part of the earth is immune, and its consequences can be devastating. Second, if the object 'like a great mountain' is to be linked to Rome, then it is a message both of destruction and self-destruction: the mountain is destroyed, but so also is the sea and much of its life. Third, nothing is so permanent, nothing is so immutable, that it cannot be damaged or destroyed if the circumstances are severe enough. This is a fragile earth and we need to live in accordance with God's will for it.

Sadly, these themes are not unique to Revelation or even to a religious milieu. The following passage is an appeal made back in 1990 by scientists hoping to alert religious leaders to the environmental crisis:

We are now threatened by self-inflicted, swiftly moving environmental alterations about whose long-term biological and ecological consequences we are still painfully ignorant ... We are close to committing – many would argue we are already committing – what in religious language is sometimes called Crimes against Creation ... Mindful of our common responsibility, we scientists ... urgently appeal to the world religious community to commit itself, in word and deed, and as boldly as is required, to preserve the environment of the Earth. As scientists, many of us have had profound experiences of awe and reverence before the universe. We understand that what is regarded as sacred is more likely to be treated with care and respect. Our planetary home should be so regarded. Efforts to safeguard and cherish the environment need to be infused with a vision of the sacred.[21]

Key message

While vast, the sea is not invulnerable. This is something that is made abundantly clear by a number of biblical passages that recognize the impact on the sea of a variety of forces, ranging from the wind, to large creatures, or the action of God. However, the Bible also perceives a connection between human sin and wider environmental devastation. Within the Bible, this is not bound up with present-day environmental issues, which its authors could not anticipate. However, its recognition of the interdependence of all aspects of creation and the profound effects of human behaviour on the wellbeing of the earth – though construed differently from modern scientific conceptualizations and often bound up with notions of divine judgement that we might find challenging today – has a new-found resonance at the present time.

Human activity now has an impact on the oceans on a scale that was unimaginable in the past. At one time the vastness of the sea could cope with what human beings threw at it, literally in the case of waste disposal, but no longer. Driven by the increasing population of the earth and the consequent demand for resources and energy, even the immense sea has become vulnerable to our human influence. No longer can we safely ignore our impact on the ocean and the creatures living in it, seeing it as we once did as the proverbial mere drop in that ocean. We are polluting the ocean both directly, as we dispose of things like plastics, and indirectly, as our use of fossil fuels increases the carbon dioxide levels in the atmosphere, which in turn acidifies the ocean as the carbon dioxide is absorbed by it. Additionally, the extra carbon dioxide in the atmosphere causes the earth to warm through the so-called 'greenhouse effect'. This in turn warms the oceans and so impacts the coral reefs, leading to coral bleaching (death of the coral) and consequently damaging the biodiversity of this complex but delicate ecosystem.

Challenge

To satisfy our (largely Western, though increasingly copied across the planet) desire for more 'stuff' requires more natural resources and more energy. More 'stuff' means more waste of which to dispose, and more energy requires, in our present global economy, more use of fossil fuels (coal, gas and oil). There are consequences for the ocean and the creatures living in it as a result.

While there are many types of pollution causing problems in the ocean, perhaps the most insidious is the case of microplastics – as noted above – which affects marine life severely. These are passed up the food chain and ultimately will be 'reaped' by us as we consume the fish that we have poisoned. The challenge therefore is how to reduce our use of plastics in order to cut down this impact on oceanic life.

A greater use of fossil fuels leads to increased carbon dioxide levels in the atmosphere, which in turn leads to both ocean warming and acidification as the carbon dioxide is absorbed by it. As noted earlier, both the warming and the acidification affect life in the oceans, particularly the phytoplankton at the base of the oceanic food chain and coral reefs. The challenge therefore is to cut down on our use of fossil fuels.

Reflection and discussion

It is worth pausing to consider the enormity of ocean pollution, its causes and its consequences. Human beings inhabit only part of the 29 per cent of the earth that is land, and actually populations are spread very thinly in many areas of the world: the polar regions, deserts, dense jungle, remote mountains, swamps. It is extraordinary (and actually quite frightening) to imagine that we, who are so incredibly small compared to the great depth and breadth of the sea, can somehow have such a significant impact on the world's greatest ecosystem and its inhabitants. More worry-ingly still, the nature of tides and currents means that nothing dropped into the sea remains undispersed. Hence although the amount of pollution that must be poured into the sea before it has a measurable impact is significant, once introduced into the ocean, there is very little we can do about it.

We discussed above the close connection in Hebrew think-ing between actions and their consequences, sin and judgement, so that harmful consequences of wrong behaviour seem almost to flow organically from the initial act, just as reaping follows sowing. There is inevitability about it, built (according to this per-spective) into the just structures of the world as determined by God himself. Judgement and blessing are, of course, personally brought about by God, just as he is perceived as being directly involved in the day-to-day running of the world and its natural processes, its tides and seasons, day and night, rain and sunshine. We do see direct divine intervention in the Bible – and sometimes, as in the book of Revelation, this action can be extreme – but at the same time causality is also human, stemming from our right or wrong actions.

- Does this way of understanding judgement and blessing fit with your reading of the Bible, and the Old Testament in particular?
- Is it a helpful model for today?
- Is it even 'just' according to our modern perceptions of what is fair and right?
- If we must reap what we sow, what implications does this have for our behaviour?
- Many biblical passages suggest that human actions can have an impact far beyond our own particular species. Often, biblical examples of this do not seem to be rooted in ideas of causality that we can recognize as operating through scientific or social processes. Nonetheless, in our present situation, examples in the ecological sphere seem more abundant and more deep-rooted than we previously recognized. Does trusting in God in this situation mean that all will be well regardless of our behav-iour, or does it rather suggest that God is just and consistent and that we ought to live within the parameters set for us?

Action

There are two immediate things that we can do. The first is concerned with our practices regarding waste disposal. The now familiar mantra of 'reduce, reuse, recycle' should be our watchword or guiding light. Many organizations are now seeking to reduce their waste disposal to zero by implementing projects that aim to recycle all their waste. While as individuals it is perhaps more difficult to do this, it is still possible to recycle much of our household waste, and most local authorities (at least in the UK) provide the facilities to do so. The question is: are we doing it, and can we do more?

The second thing that we can do is to reduce our reliance on fossil fuels. This might seem like too big a challenge and on one level it is. Governments and public and private organizations and companies need to respond to this challenge too. However, that should not become an excuse for inaction on our part. We all have a role to play. We can downsize our car (or replace it with a more energy-efficient one), or walk, or cycle, or use public transport such as buses and trains. We can insulate our houses and turn down the heating a notch on the thermostat and wear a jumper (or even two) rather than just being in shirtsleeves. We can use low-energy light bulbs, turn off electric appliances when not in use, and consider whether it might be practical to install solar panels or make other changes to increase energy efficiency. Energy used in food production and transport is a significant contributor to global warming, so reducing food waste is another important aspect of limiting carbon emissions.

In fact, since nearly everything we do has an environmental impact, we should make it a habit to consider this in respect of our small actions throughout the day. For example, from where is my food sourced? Which ingredients have a lower carbon impact? How might I prepare it so as to use less energy? Can I use washing-up liquid that will have a lesser effect on the environment than conventional detergents, etc.? Often it is difficult to form a judgement as to the environmental impact of our choices. However, there are numerous online sources that can help us take action to save energy and calculate our carbon footprint.[22] While such changes will not solve the problem, they will contribute towards the solution – even if only in a small way. Surely it is better to be part of the solution than part of the problem?

Notes

1 'Pre-industrial' refers to periods prior to the Industrial Revolution of the late eighteenth to early nineteenth century when the use of fossil fuels increased rapidly.

2 In addition to degradation of larger pieces of plastic, microplastics are also manufactured plastics of microscopic size used for industrial and domestic purposes, such as in air blasting media, facial cleaners and cosmetics, and in medicine for drug delivery.

3 See, for example, Job 38.8–11, Psalm 104.9, Proverbs 8.29 and Jeremiah 5.22. This theme is discussed in more detail in Chapter 6.

4 An interesting reflection of this distinction is in the reference to the 'Sea' of Galilee in three of the Gospels, whereas Luke, perhaps the most Greek of the evangelists, consistently describes it as a 'lake'.

5 The depletion of these and other inland seas or lakes is graphically seen in aerial photographs: www.ibtimes.co.uk/world-water-day-2016-before-after-images-earths-disappearing-lakes-seas-1550383.

6 See Chapter 3 for more on phytoplankton. Note that some phytoplankton can luminesce (that is, emit light) when the water around them is disturbed. This is thought to be a defence mechanism against predators.

7 Or 'splendour' or 'majesty': the NRSV 'beauty' does not quite capture the sense of honour that such a head commands.

8 Proverbs 20.29.

9 Proverbs 16.31.

10 Genesis 15.15, 25.8; Judges 8.32; 1 Chronicles 29.28.

11 Aeschylus, *Persians*, 599.

12 Sophocles, *Oedipus Tyrannos* (*Oedipus the King*), 1527.

13 Euripides, *Hecuba* 118.

14 Sophocles, *Electra* 733.

15 Euripides, *Ion* 60.

16 This is again probably another reference to the Euphrates: note the parallelism between Egypt and Assyria that follows.

17 This is the translation found in the NRSV in, for example, Deuteronomy 15.9, 23.21–22, 24.15. Clearly the consequence of sin is in mind in these passages, though the concept of 'punishment' could alternatively (or additionally) be expressed here.

18 This seems to be the meaning, for example, in Leviticus 20.20, 24.15, Numbers 9.13 (translated in the NRSV as 'consequences for the sin'), and

Lamentations 3.39 ('punishments of their sins' in the NRSV). Perhaps the best-known example is in Isaiah 53.12, where it is said that the servant of the LORD 'bore the sin of many', clearly signifying that he took on himself the consequences for others' sin:

Therefore I will allot him a portion with the great,
  and he shall divide the spoil with the strong;
because he poured out himself to death,
  and was numbered with the transgressors;
yet he bore the sin of many,
  and made intercession for the transgressors.

19 The relevant part of the verse reads: 'For they sow the wind, and they shall reap the whirlwind.'

20 Historically, Christians under persecution have found the book of Revelation more relevant and comprehensible than perhaps we do, which is unsurprising as it was written to Christians who were suffering oppression.

21 This was published in *Preserving and Cherishing the Earth: An Appeal for Joint Commitment in Science and Religion* (1990). The statement emanated from the Global Forum, Moscow: National Religious Partnership for the Environment, January 1990, and is available online at The Forum on Religion and Ecology (http://fore.research.yale.edu/publications/statements/preserve/).

22 www.climatecare.org/climate/low-carbon-living/.

8

Economics, Hubris and Human Community: Travel and Trade on the Sea

Sailing into the blue

Both the Old and New Testaments describe sea voyages of various kinds, some of which are related to trade and some undertaken simply to travel from one place to another. For example, Paul journeyed to Rome on a ship that was also carrying cargo (Acts 27.6, 10; being an Alexandrian ship, it was probably carrying grain from Egypt). According to 1 Kings 9–10, during the reign of Solomon ships sailed from Ezion Geber (modern-day Eilat – the southernmost point of Israel on the Gulf of Aqaba), bringing back exotic cargoes including gold from Ophir, silver, ivory, apes and peacocks (1 Kings 9.26–28, 10.22).[1] Often early traders such as these were literally sailing off into the blue – into unknown and uncharted seas – seeking to make their fortunes.[2] It is only since the nineteenth century – comparatively recently – that humans have completely charted the seas and that element of sailing into the blue, into the unknown, to trade has completely vanished.[3]

The poem *Cargoes* by John Masefield captures some of the flavour of the biblical, as well as more recent, trading voyages on the sea, with the first verse clearly echoing the exotic goods listed in connection with the splendour of Solomon's reign:

Quinquireme of Nineveh from distant Ophir,
Rowing home to haven in sunny Palestine,
With a cargo of ivory,

And apes and peacocks,
Sandalwood, cedarwood, and sweet white wine.

Stately Spanish galleon coming from the Isthmus,
Dipping through the Tropics by the palm-green shores,
With a cargo of diamonds,
Emeralds, amethysts,
Topazes, and cinnamon, and gold moidores.

Dirty British coaster with salt-caked smoke stack,
Butting through the Channel in the mad March days,
With a cargo of Tyne coal,
Road-rails, pig-lead,
Firewood, ironware, and cheap tin trays.

The poem encapsulates both the exotic and the prosaic aspects of trade carried out on the seas. Today huge quantities of goods are shipped round the world and the ocean is a key part of the global economy. Here we will explore trade and travel on the sea and the economic significance of the ocean from a modern and a biblical perspective.

Blue economy

With the ever-expanding demand for food and resources, together with the consequent requirement to transport goods over large distances, there has been a new focus in recent years on the so-called 'blue economy'. The blue economy is basically the use of the world's oceans for economic purposes. Of course, there is nothing new about this. Since prehistoric days humans have obtained food from the oceans through fishing, and even travelled long distances across the ocean in very simple and seemingly flimsy craft – defying the elements that could so easily destroy them. Both fishing and travel led to trade, and therefore economic use of the oceans.

In biblical times we can think of the Bronze Age Phoenician economy, built on the trade across the Mediterranean of 'royal' purple cloth manufactured using dye from the murex shellfish. Later, grain ships plied their trade between Egypt and Rome, the heart of the Roman Empire. With the discovery of the 'New World' in 1492 by Columbus, European ships began to cross the Atlantic, taking settlers in one direction and bringing gold and other goods back in the other. Sadly, this developed into the transport of slaves from Africa, as a less than proud part of European history. More recently still, the British Empire was sustained through travel and trade across the global ocean and 'Britannia rule the waves' was the accompanying refrain.

Today huge container ships cross the oceans carrying all the consumer goods and foods that modern (First World) society thinks it requires: cars, washing machines, cookers, computers and exotic fruits and vegetables, to name but a few. By 2006, international seaborne freight constituted 90 per cent of global trade by volume.[4] Additionally, the oil that is (literally and metaphorically) required to fuel our society crosses the sea in large tankers. All this drives a global economy worth hundreds of billions of dollars. If we add to this the tourism that is associated with the sea – cruise liners, beach holidays on exotic islands and coastlines, sailing, surfing and so on – the role of the ocean in the global economy is further magnified.

The term 'blue economy' has two meanings. The first is simply the economic activity that is associated with the earth's oceans and seas. The second is a meaning equivalent to the term 'green economy', but applied to the ocean – that is, a sustainable use of the oceans for economic purposes. It is clear that humanity is not developing a blue economy in the second sense at this time, though there are initiatives to try to do so.

Blue growth

Associated with the blue economy is the idea of 'blue growth' – that is, the expansion of the blue economy. This is underpinned by the idea, more broadly held, that economic growth is fundamentally a good goal that society should pursue. Of course, this idea has been challenged for a variety of reasons, not least by the fact that the earth's resources – whether oceanic or terrestrial – are finite, whereas sustained economic growth seems to demand ever more resources.

For some, though, blue growth means sustainable growth in the use of the oceans for economic purposes. Unfortunately, humanity seems already largely to have over-exploited the more easily available resources such as fish and crude oil reserves below the seabed. Whether we can 'turn back the clock' and achieve sustainable use of the oceans for economic growth is a moot point. The desire to have sustainable development goals for the ocean, just as for the rest of the environment, has led to the formulation of principles that might help to achieve this.[5] These include providing equitable access to ocean resources and ensuring that neither pollution nor the harvesting and extraction of living or mineral resources negatively affects the ocean ecosystem. In addition, the aim is to help coastal communities, particularly in poorer countries who rely on the oceans, to increase their resilience and develop more sustainable economies. In this way it is hoped that the wealth of the oceans can be of benefit to more of humanity than it currently is.

Blue wealth

One can think of blue wealth in two ways. One is associated with the blue economy and blue growth, wealth in the sense of growing rich from oceanic resources. However, there is also the wealth of human experience, ideas and culture that has been exchanged between societies across the earth as men and women have crossed the world's oceans. This wealth is perhaps less tangible but nevertheless of great worth. As we encounter those who are different from us, our lives can be enriched through that interaction. As well as being a barrier that divides, the oceans can also be seen as a means of connecting humans around the planet. This has long been the perspective of those who have lived on the scattered islands across the vast Pacific Ocean, yet have been linked in terms of culture and trade by the ocean, which allows travel between the islands.

In the UK, our view of the sea can be a little more ambivalent and circumstantial. At present, many people seem to want to see the English Channel as dividing us from the rest of the continent of Europe and providing a barrier that keeps us safe and secure.[6] This insular mentality can lead to a diminished appreciation of the cultural and economic wealth that comes through interactions with our neighbours on the European mainland. In the past, of course, in the days when 'Britannia ruled the waves', people in the UK probably regarded the ocean more positively, as it linked the far-flung reaches of the British Empire. At other times, such as when Sir Francis Drake defeated the Spanish Armada and when Hitler's armies failed to invade the UK, the sea has provided a welcome barrier against invaders.

Financially, sources of blue wealth include those aspects of the oceans that are of direct economic relevance – like fisheries, aquaculture, offshore oil and gas mining, shipping, tourism and, in the future perhaps, deep seabed mining for minerals. Although not strictly an ocean activity, the dismantling of disused ships beached on the shores of economically poorer countries is another source of wealth for some (though not for those dismantling the ships, of course), since scrap metal is a valuable commodity. There is no doubt that some people have grown wealthy by exploiting ocean resources. The question is whether this is sustainable or even just.

'Just blue'

A vivid image that has recently impacted people is the sight of refugees trying to cross the Mediterranean to Europe in boats that might be considered less than seaworthy. Many people have lost their lives trying to make the crossing from North Africa to Europe, and even on the shorter crossing from Turkey to Greece. The sea presents a means of escape, albeit a dangerous one.

As noted above, issues of justice arise in how to respond to the refugees arriving on our shores from across the sea. Likewise, the whole blue economy, blue growth and blue wealth agenda raises questions of equity. Who controls the oceans' resources? Who gains from them? Who takes responsibility for their wise use to the benefit of the many rather than just the few? The challenge is: is our response to these issues a just one?

'Just blue' can express how we feel about some of the exploitation of people and resources associated with the earth's oceans. It can express too our emotional response to the ongoing refugee crisis that is driving people to risk their lives crossing the seas in unsuitable boats. Perhaps it also expresses how God feels when he looks on the same injustices affecting people and the planet: a sense of sadness that his good world (Genesis 1) has come to this because the people he asked to look after it – us – are failing to do so.

'Just blue' also speaks of the beauty of the ocean, as anyone who has looked on a deep blue sea sparkling in the sunshine will acknowledge. One of us (Meric) is a 'blue water' oceanographer, being concerned with the deep ocean where the colour of the water is less affected by other factors, unlike the so-called green waters of the coastal ocean and the brown waters of rivers and estuaries, whose colours are more altered by algal growth and sediments, respectively. Looking at photographs of the earth taken from space we are reminded that we live on a largely blue planet, sometimes referred to as a 'blue marble' floating serenely in space. So the earth is 'just blue' too.

Sea travel in the Bible

As noted earlier, it is easy to think of the sea as a barrier more than as a connector, since in a world of modern (albeit sometimes congested) roads or railway lines, it is more time-consuming and awkward to travel the 20.6 miles across the Channel from Dover to Calais than an equivalent distance over land. Certainly, this attitude to the sea as a barrier is also reflected in the Bible. When the people of Israel made their Exodus from Egypt, the defining moment of their exit was the miraculous crossing of the Red Sea – a clear (and seemingly insuperable) barrier that marked their departure from Egypt. Likewise, the entry to the Promised Land involved a parallel event – the crossing of the River Jordan, which again, according to Joshua 3.16–17, was stopped like the sea as the people passed over. In the Old Testament, the ideal boundary of the Promised Land at its fullest extent was 'from the sea to the river' (that is, from the Mediterranean to the Euphrates),[7] and the sea (or a river, even a small one like the Jordan) is often seen as a boundary marker between national or tribal areas.[8]

However, in terms of ease of travel over long distances, in the ancient world the sea was to be preferred over the land. It made much better sense for Jesus and his disciples to cross the Sea of Galilee by boat than to walk round to the other side. Paul on his missionary journeys took the Christian message around the northern Mediterranean coast to Rome, but it took much longer to percolate inland. And in the book of Jonah, when the reluctant prophet wanted to flee as far and as quickly from God as possible, he immediately made for the coast to board a ship to Tarshish. This port (as we saw before) was probably on the Atlantic coast of Spain – an unthinkable distance to cover by any other means within that ancient context (especially, one might think, if you were hoping not to be overtaken by God before you got very far!).

Maritime trade in the Bible

You only have to look at a container ship to gain an impression of its immense capacity as compared to that of even a very large lorry. In fact, where speed is not an important factor, it can still make sense to transport heavy cargoes such as landfill or gravel by boat on the inland waterways rather than by lorry, and capacity is an important aspect of this.[9]

In the ancient world, the land-borne alternative to sea freight was a caravan of pack animals – or, more accurately, a series of caravans. Each of these would travel the sometimes very long distance between given exchange points, so that they undertook only a section of the journey before the products they carried were traded again. As a result, the place of origin of many precious commodities such as spices was often unknown to the end-user, and purchase prices reflected the considerable series of merchant transactions behind their transport. The challenges of weight and bulk when reliant on pack animals (especially on difficult terrain) also restricted the breadth of products that could be transported in this way over any great distance. Spices and precious stones or gold and, in New Testament times, even Chinese silk, could be traded as relatively compact and high-value products, but bulk goods like grain were much less likely to be exchanged far from their place of origin.

Maritime trade therefore presented a viable alternative, enabling the transport of larger quantities and more varied goods over long distances. This may have reduced the long supply-chains, but it was also costly: besides the demands of boat-building and maintenance, as well as of sustaining a crew of oarsmen, it was also high-risk in terms of the probability of shipwreck or damage to the boat or cargo, and seasonally dependent. Given the conflicting demands of rowability and accommodation of supplies and other needs for the crew over a long journey, as compared with the stowage of saleable goods, it was still an expensive process. Nonetheless, ancient shipwrecks provide us with evidence of the transportation within the Mediterranean of food products including oil, wine and garum (a fish sauce much loved by the Romans), metals such as iron, tin or silver, and also cloth, including some dyed with purple murex, to which the Phoenicians owed their wealth.

Conflicted desires

Collectively, passages in the Bible concerning sea-trade beautifully exemplify a very human ambivalence towards it. On the one hand, the great wealth that it brought, and the power and cultural superiority that the possession of a merchant fleet embodied, were objects of pride and desire. On the other hand, the arrogance and excessive wealth of mercantile peoples also excited horror, if also wonder, at the opulent and exotic products that could be imported – and maybe also more than a touch of envy.

Pride and desire

Possibly one of the most transparent expressions in literature of the desirability of wealth and power is found in 1 Kings 9.26—10.29 and its parallel in 2 Chronicles 8.17—9.28 and in 2 Chronicles 1.14–17. These passages are rather long to reproduce here in full, but this is an abbreviated version of that in Kings:

> [26]King Solomon built a fleet of ships at Ezion-geber, which is near Eloth on the shore of the Red Sea, in the land of Edom. [27]Hiram [the king of Tyre, a Phoenician city] sent his servants with the fleet, sailors who were familiar with the sea, together with the servants of Solomon. [28]They went to Ophir, and imported from there four hundred and twenty talents of gold, which they delivered to King Solomon.
>
> ...
>
> [10.2][The queen of Sheba] came to Jerusalem with a very great retinue, with camels bearing spices, and very much gold, and precious stones ... [10]Then she gave the king one hundred and twenty talents of gold, a great quantity of spices, and precious stones; never again did spices come in such quantity as that which the queen of Sheba gave to King Solomon.
>
> [11]Moreover, the fleet of Hiram, which carried gold from Ophir, brought from Ophir a great quantity of almug wood and precious stones ...
>
> [14]The weight of gold that came to Solomon in one year was six hundred and sixty-six talents of gold, [15]besides that which came from the traders and from the business of the merchants, and from all the kings of Arabia and the governors of the land ... [21]All King Solomon's drinking vessels were of gold, and all the vessels of the House of the Forest of Lebanon were of pure gold; none were of silver – it was not considered as anything in the days of Solomon. [22]For the king had a fleet of ships of Tarshish at sea with the fleet of Hiram. Once every three years the fleet of ships of Tarshish used to come bringing gold, silver, ivory, apes, and peacocks.
>
> [23]Thus King Solomon excelled all the kings of the earth in riches and in wisdom ... [25]Every one of them brought a present,

objects of silver and gold, garments, weaponry, spices, horses, and mules, so much year by year.

²⁶Solomon gathered together chariots and horses; he had fourteen hundred chariots and twelve thousand horses, which he stationed in the chariot cities and with the king in Jerusalem. ²⁷The king made silver as common in Jerusalem as stones, and he made cedars as numerous as the sycamores of the Shephelah. ²⁸Solomon's import of horses was from Egypt and Kue,* and the king's traders received them from Kue at a price. ²⁹A chariot could be imported from Egypt for six hundred shekels of silver, and a horse for one hundred and fifty; so through the king's traders they were exported to all the kings of the Hittites and the kings of Aram. (1 Kings 9.26–10.29)

[*The location of Kue is debated, but somewhere near Egypt or in Anatolia seem the most likely possibilities.]

The passage clearly sets out to describe 'Solomon in all his glory' in the most superlative terms possible. However, wealth, and in particular embarking on ambitious trading ventures, is understood as a prominent demonstration of the greatness of this king. Particularly interesting is the transference of power by association in a rare biblical attempt at name-dropping. Solomon is in one stroke portrayed both as having his own mercantile fleet and as being able to engage the expertise of the master of the most powerful navy in the known world, Hiram, king of Tyre, in this ambitious venture. Sheba and Ophir, both of which were most probably in Arabia, are specifically mentioned, Arabia being understood as the source of some of the most prized commodities in the ancient world: gold, precious stones, and spices such as frankincense.[10] The success of his forays into this risky, yet highly lucrative trade, are further demonstrated by the import of extraordinarily precious and exotic products: besides gold, precious stones, horses and the clearly special but now unknown almug wood, the NRSV mentions 'silver, ivory, apes, and peacocks'.[11] In addition, 'all the kings of the earth' are described as bringing him gifts of 'objects of silver and gold, garments, weaponry, spices, horses, and mules'. The whole thrust of this passage in conjuring up a realm of unimaginable wealth is nicely epitomized by the idea that silver 'was not considered as anything in the days of Solomon'.

However, this proud – and maybe defiant – depiction of extraordinary wealth, success and power is also on another level infused with pathos. The books of Kings encompass the period from Solomon until the Fall of Jerusalem in 586 BC and the end of the monarchy, at which point the city was conquered by the Babylonians and its leading citizens taken into exile. In its present form at least, 1 and 2 Kings must stem from a period from which the past was viewed with nostalgia as a lost golden age. Already in its treatment of the reign of Solomon's successor, his son Rehoboam, it records the division of the monarchy into the separate kingdoms of Judah and Israel,[12] while the fragmentation of Solomon's mini-empire is hinted at in 1 Kings 11.11–13. Thereafter, the picture is predominantly one of struggle and suffering as the small kingdoms came successively under the dominance of the greater powers of Syria, Assyria (under whom Israel fell) and Babylon (at which point, Judah, too, ceased to exist as a state).[13] The portrayal of wealth and maritime prowess, then, is not just a celebration of a glorious age, but is tinged with loss and longing, and with dreams of what might have been – or might be yet. It is the political and economic equivalent of the Garden of Eden.

There is another possible aspect to the reading of this passage. Archaeological records are notoriously open to differing interpretations, not least because of the limited extent of what has been preserved and discovered from such a distant time. In many cases, interpreting the evidence is also subject to debates over the correct dating of archaeological finds. Nonetheless, many scholars question the extent of the wealth and influence Judah had in Solomon's time, so the picture painted in Kings and Chronicles may be somewhat idealized. However, one thing we can be sure of is that irrespective of how glorious Solomon's reign was (or was not), the tone of this passage clearly reveals the allure of wealth and power, and the prestige that attends it.

Great pride and an even greater fall

Notwithstanding the claims to glory and power embodied in the chapters from Kings and Chronicles that we have just discussed, a more prevalent biblical voice is one that finds more horror than delight in the trappings of wealth and in the pride and arrogance that often attends it.

The following passage from Ezekiel has no reservation in describing in detail the often valuable and desirable items in which the merchants of Tyre traded, the craftsmanship embodied in their beautifully fitted ships, and the wealth and influence that this betrays. The extent of the Tyrians'[14] trading networks, as well as the range of the products in which they dealt, excites wonder and awe:

> ³... say to Tyre, which sits at the entrance to the sea, merchant of the peoples on many coastlands, Thus says the Lord GOD:
>
> O Tyre, you have said,
>   'I am perfect in beauty.'
> ⁴Your borders are in the heart of the seas;
>   your builders made perfect your beauty.
> ⁵They made all your planks
>   of fir trees from Senir;
> they took a cedar from Lebanon
>   to make a mast for you.
> ⁶From oaks of Bashan
>   they made your oars;
> they made your deck of pines
>   from the coasts of Cyprus,
>   inlaid with ivory.
> ⁷Of fine embroidered linen from Egypt
>   was your sail,
>   serving as your ensign;
> blue and purple from the coasts of Elishah
>   was your awning ...
>
> ¹²Tarshish did business with you out of the abundance of your

great wealth; silver, iron, tin, and lead they exchanged for your wares. ¹³Javan, Tubal, and Meshech ... exchanged human beings and vessels of bronze for your merchandise. ¹⁴Beth-togarmah exchanged for your wares horses, war-horses, and mules. ¹⁵The Rhodians ... brought you in payment ivory tusks and ebony. ¹⁶Edom ... exchanged for your wares turquoise, purple, embroidered work, fine linen, coral, and rubies. ¹⁷Judah and the land of Israel ... exchanged for your merchandise wheat from Minnith, millet, honey, oil, and balm. ¹⁸Damascus traded with you for your abundant goods ... wine of Helbon, and white wool. ¹⁹Vedan and Javan from Uzal entered into trade for your wares; wrought iron, cassia, and sweet cane were bartered for your merchandise. ²⁰Dedan traded with you in saddlecloths for riding. ²¹Arabia and all the princes of Kedar were your favoured dealers in lambs, rams, and goats ... ²²The merchants of Sheba and Raamah ... exchanged for your wares the best of all kinds of spices, and all precious stones, and gold. ²³Haran, Canneh, Eden, the merchants of Sheba, Asshur, and Chilmad traded with you ²⁴... in choice garments, in clothes of blue and embroidered work, and in carpets of coloured material, bound with cords and made secure ... ²⁵The ships of Tarshish travelled for you in your trade.

So you were filled and heavily laden
   in the heart of the seas. (Ezekiel 27.3–25)

Not all the places named here are identifiable, but it seems to encompass much of the then known world, indicating the astonishing reach of Phoenician trading networks.¹⁵ Her 'borders are in the heart of the seas' (Ezekiel 27.4): her trading enterprises are a truly international affair, and all the world seeks to do business with her and seems to be held in her thrall. Nor is the beauty of the great ship Tyre, or the great skill involved in her undertakings, underplayed. She is said to be 'perfect in beauty' (vv. 3–4, 11) and the abundance of her wealth and goods is repeatedly stressed (vv. 12, 16, 18).

Yet Ezekiel does not stand in awe of Tyre. Rather, he perceives that the power and magnificence of this city-state had engendered an extraordinary level of arrogance and self-confidence. If you look again at the first verse (v. 3) of the passage just quoted, it begins, 'O Tyre, you have said, "I am perfect in beauty."'

The perfection of her beauty may not be denied, as it is mentioned again in vv. 4, 11 and 28.12 – unless, of course, the claim attributed to Tyre is quoted again sarcastically. However, the arrogance that this reveals is inexcusable.[16] Thus, Ezekiel 27 begins with the words: 'The word of the LORD came to me: Now you, mortal, raise a lamentation over Tyre ...' (Ezekiel 27.1–2). The description of the great ship Tyre and the astonishing array of nations wishing to do business with her that we have just considered is effectively a eulogy uttered after her certain death.

Ezekiel then continues in 27.26–36 by describing that death, which is envisaged as the wrecking of the great ship Tyre, together with all her riches and merchandise, her merchants and warriors and all who are in her: all of them 'sink into the heart of the seas' (vv. 26–27). The passage describes the distress and shock at her unexpected passing, and the elaborate rituals of grief engaged in by the mariners who stand by the shore weeping, crying aloud, shaving their heads and donning sackcloth and ashes, while uttering a lament over her (vv. 28–32).

However, besides detailing the wrecking of the great ship Tyre, Ezekiel imagines the international reverberations of her fall. As he appreciates:

> When your wares came from the seas,
>   you satisfied many peoples;
> with your abundant wealth and merchandise
>   you enriched the kings of the earth. (Ezekiel 27.33)

The loss of Tyre has a huge impact, not least on the kings (its chief beneficiaries) themselves. No one had imagined that this great trading people could meet an end such as this, and news of its demise meets with widespread shock, distress and fear – and apparently ultimately derision:

> [35]All the inhabitants of the coastlands
>   are appalled at you;
> and their kings are horribly afraid,
>   their faces are convulsed.

³⁶The merchants among the peoples hiss at you;
    you have come to a dreadful end
    and shall be no more forever. (Ezekiel 27.35–36)

The chapter therefore powerfully reveals the interconnectedness of peoples through trade, the fears and aspirations of those in power, and their common focus on wealth and status. However, it also reflects not only the pride of Tyre, but the common acceptance of her invincibility, the failure of which caused profound shock well beyond her boundaries.

Godlike Tyre?

Ezekiel unpacks the reasons for the spectacular fall of Tyre in the following chapter, 28. Addressing himself to the city-state's ruler, the prophet accuses him of such overweening pride that he even considers himself godlike both in his confident location in the midst of the sea and in his wisdom. The source of Tyre's splendour and wisdom, however, is its wealth and extensive trading connections. This nation's extraordinary command over the sea and its hold over its weaker trading partners had created a compelling fallacy of power:

    ¹The word of the LORD came to me: ²Mortal, say to the prince
      of Tyre, Thus says the Lord GOD:
      Because your heart is proud
        and you have said, 'I am a god;
      I sit in the seat of the gods,
        in the heart of the seas',
      yet you are but a mortal, and no god,
        though you compare your mind
        with the mind of a god.
      ³You are indeed wiser than Daniel;
        no secret is hidden from you;
      ⁴by your wisdom and your understanding
        you have amassed wealth for yourself,
     and have gathered gold and silver
        into your treasuries.

⁵By your great wisdom in trade
  you have increased your wealth,
  and your heart has become proud in your wealth.
⁶Therefore thus says the Lord GOD:
 Because you compare your mind
  with the mind of a god,
⁷therefore, I will bring strangers against you,
  the most terrible of the nations;
 they shall draw their swords against the beauty of your
  wisdom
  and defile your splendour ...
⁸They shall thrust you down to the Pit,
  and you shall die a violent death
  in the heart of the seas.
⁹Will you still say, 'I am a god,'
  in the presence of those who kill you,
 though you are but a mortal, and no god,
  in the hands of those who wound you?
¹⁰You shall die the death of the uncircumcised
  by the hand of foreigners;
  for I have spoken, says the Lord GOD. (Ezekiel 28.1–10)

Ezekiel treads a fine line between openly acknowledging Tyre's genuine wealth, wisdom and beauty, and at the same time reprimanding her for imagining that this equated to godlike power. The NRSV translation of v. 2 reads 'you are but a mortal and no god', and this summarizes Ezekiel's rebuke to Tyre very well. In fact, the Hebrew does not specifically refer to mortality, since it uses the word *ādām*, 'human being', which is so familiar to us from Genesis 2. However, the Garden of Eden story in Genesis emphasizes that the *ādām* is made from the *ādāmāh*, 'earth' or 'dust'.¹⁷ This is expressed initially in the description of the making of the first human being from the dust in the ground (Genesis 2.7), and then of course recurs in the curses at the end ('You are dust and to dust you shall return', Genesis 3.18). It may well be therefore that this transient aspect of human nature is indeed in his mind in the use of the word *ādām* here in Ezekiel 28.2.

In any case, Tyre will certainly discover what it is to be mortal, since Ezekiel talks on various levels of the death of the prince and his nation, of being killed and wounded at the hands of foreigners (Ezekiel 28.7–10) and 'thrust down to the Pit' (that is, the underworld, v. 8). The idea of dying in the heart of the seas is doubled-edged: Tyre was an island nation, so in a sense would meet its demise in just such a location. But dangerous waves and the experience of being submerged in deep water can also provide a metaphor for death: if you look at the psalm in Jonah 2, the imagery works on both literal and metaphorical levels: the waters are those of the sea and also of the underworld.[18] Finally, proper burial practices were a vital part of managing death in the ancient world, but Tyre will be denied the proper ceremonies and procedures, which would probably be understood to inhibit proper entry into the world of the dead.

If immortality is one key aspect of what it is to be divine, another one, in the Old Testament at least, is wisdom. This is seen in Genesis 2—3, in the story of Adam and Eve, by the way the tree of life is complemented by the tree of the knowledge of good and evil: the fruits of both are forbidden to human beings because they represent qualities that are the exclusive preserve of God. Claims to both attributes seem to be at issue here in the condemnation of Tyre.

Nonetheless, just as Ezekiel himself seemed to acknowledge that Tyre was indeed 'perfect in beauty', so he also here agrees that its 'prince' (embodying the nation itself) is wiser than a proverbially wise man, Daniel: he is wealthy precisely because of his wisdom and understanding. Tyre truly is impressive in its wealth and wisdom and beauty – and Ezekiel does nothing to deny the advantages that wealth can bring, nor the attractiveness of having the best money can buy, with power and influence to boot. At the same time, its ruler's pretensions to a divine level of wisdom, comparing his mind (in Hebrew, literally 'heart', which was thought to be the seat of thinking) to that of a god, are deserving of the highest level of condemnation. Wisdom has created great wealth, and this success has brought excessive pride, worthy of an even greater fall.

It is not clear whether the king of Tyre may actually have claimed to have been a god. It is certainly possible, since similar claims were made, for example, by Egyptian pharaohs, each of whom were believed to be the embodiment of Horus and a son of Re. However, Ezekiel may simply be observing a life lived by a man who felt totally in command of his destiny. He was wise, powerful, influential, rich – and could live his life without reference to God, believing in the chimera of his own self-determination and in the apparent invincibility and permanence of Tyre.

The corrupting power of wealth

There may also be a further aspect to Ezekiel's condemnation of Tyre, though, and this is in the ethical conduct of trade. Ezekiel 27 already perceptively observes how:

> 33When your wares came from the seas,
>   you satisfied many peoples;
> with your abundant wealth and merchandise
>   you enriched the kings of the earth. (Ezekiel 27.33)

It is those who are already rich and powerful who profit from such trade. That may not necessarily of itself provide grounds for condemnation. However, as the passage continues, it suggests that violence and unrighteousness went hand-in-hand with Tyre's mercantile activity.

Ezezkiel 28.11–19 looks obscure, but it is referring back to the Garden of Eden tradition. The one we are most familiar with is of Adam as 'fallen', the sinner who was cast out of the Garden. However, there is another tradition that sees the first human as glorious – a tradition that eventually found its focus in Christ as the new Adam, the man of glory, as contrasted with the first Adam who sinned and brought death into the world. It is this idea of Adam as an ideal man to which Ezekiel refers – although as the passage develops, we see that here these promising beginnings are also disappointed:

> 11Moreover, the word of the LORD came to me: 12Mortal, raise a lamentation over the king of Tyre, and say to him, Thus says the Lord GOD:

You were the signet of perfection,
full of wisdom
and perfect in beauty.
¹³You were in Eden, the garden of God;
every precious stone was your covering,
carnelian, chrysolite, and moonstone,
beryl, onyx, and jasper,
sapphire, turquoise, and emerald;
and worked in gold were your settings
and your engravings.
On the day that you were created
they were prepared.
¹⁴With an anointed cherub as guardian I placed you;
you were on the holy mountain of God;
you walked among the stones of fire.
¹⁵You were blameless in your ways
from the day that you were created,
until iniquity was found in you.
¹⁶In the abundance of your trade
you were filled with violence, and you sinned;
so I cast you as a profane thing from the mountain of God,
and the guardian cherub drove you out
from among the stones of fire.
¹⁷Your heart was proud because of your beauty;
you corrupted your wisdom for the sake of your
splendour.
I cast you to the ground;
I exposed you before kings,
to feast their eyes on you.
¹⁸By the multitude of your iniquities,
in the unrighteousness of your trade,
you profaned your sanctuaries.
So I brought out fire from within you;
it consumed you,
and I turned you to ashes on the earth
in the sight of all who saw you.
¹⁹All who know you among the peoples
are appalled at you;

you have come to a dreadful end
and shall be no more forever. (Ezekiel 28.11–19)

Here sin comes through trade – not because trade is necessarily condemned *per se*, but because it was conducted with violence: 'In the abundance of your trade you were filled with violence, and you sinned; so I cast you as a profane thing from the mountain of God' (Ezekiel 28.16). It seems almost as if there is something about the scale of Tyre's exchanges that itself leads to corruption. However, to the unrighteousness of Tyre's trade (v. 18) are also joined the sin of pride (v. 17a) and the accusation that 'you corrupted your wisdom for the sake of your splendour' (v. 17b). Presumably, Tyre used its skills and wisdom for self-serving ends – or in devising inappropriate means to achieve those ends – and is thus deserving of condemnation.

The intriguing thing about these two chapters from Ezekiel is that it does nothing to deny the wonders that wealth can bring, with beautiful and luxurious products, status and power. The riches of Tyre and the city's own attributes are presented in all their glory. However, it is also equally explicit about the pitfalls of such privilege: the pride, arrogance, violence and corruption that can attend the undiluted pursuit of such wealth.

Iniquitous trade

Ezekiel's voice finds a further echo in the New Testament, in Revelation 18, which bears more than a passing resemblance to the oracle against Tyre in Ezekiel 27—28, but the condemnation is now directed against 'Babylon', aka Rome:

> [1]After this I saw another angel coming down from heaven, having great authority; and the earth was made bright with his splendour. [2]He called out with a mighty voice,
> 'Fallen, fallen is Babylon the great! ...
> [3]For ... the kings of the earth have committed fornication
> with her,
> and the merchants of the earth have grown rich from the
> power of her luxury.'

⁴Then I heard another voice from heaven saying,
‘... ⁷As she glorified herself and lived luxuriously,
    so give her a like measure of torment and grief.
Since in her heart she says,
    'I rule as a queen;
I am no widow,
    and I will never see grief',
⁸therefore her plagues will come in a single day –
    ... for mighty is the Lord God who judges her.'

⁹And the kings of the earth, who committed fornication and
lived in luxury with her, will weep and wail over her when they
see the smoke of her burning; ¹⁰they will stand far off, in fear of
her torment, and say,
    'Alas, alas, the great city,
    Babylon, the mighty city!
For in one hour your judgement has come.'

¹¹And the merchants of the earth weep and mourn for her, since
no one buys their cargo anymore, ¹²cargo of gold, silver, jewels
and pearls, fine linen, purple, silk and scarlet, all kinds of scented
wood, all articles of ivory, all articles of costly wood, bronze,
iron, and marble, ¹³cinnamon, spice, incense, myrrh, frankincense, wine, olive oil, choice flour and wheat, cattle and sheep,
horses and chariots, slaves – and human lives.
    ¹⁴'The fruit for which your soul longed
        has gone from you,
    and all your dainties and your splendour
        are lost to you,
        never to be found again!'
¹⁵The merchants of these wares, who gained wealth from her,
will stand far off, in fear of her torment, weeping and mourning
aloud,
    ¹⁶'Alas, alas, the great city,
        clothed in fine linen,
            in purple and scarlet,
        adorned with gold,
            with jewels, and with pearls!

[17]For in one hour all this wealth has been laid waste!'
And all shipmasters and seafarers, sailors and all whose trade is
on the sea, stood far off [18]and cried out as they saw the smoke
of her burning,
   'What city was like the great city?'
[19]And they threw dust on their heads, as they wept and mourned,
crying out,
   'Alas, alas, the great city,
     where all who had ships at sea
     grew rich by her wealth!
   For in one hour she has been laid waste ...'
[21]Then a mighty angel took up a stone like a great millstone and
threw it into the sea, saying,
   'With such violence Babylon the great city
     will be thrown down,
     and will be found no more;...
[23]... for your merchants were the magnates of the earth,
     and all nations were deceived by your sorcery.
[24]And in you was found the blood of prophets and of saints,
     and of all who have been slaughtered on earth.'
                  (Revelation 18.1–24)

On a literal level, the presentation of Rome as a great maritime
power is just as apt as it was to Tyre: the movement of trading
ships was an important aspect of the Roman Empire and vital to
keep it fed and supplied with all necessary goods, as well as with
more luxurious products. Indeed, so important was it that the
Romans were the first people to have a marked effect on piracy
in the Mediterranean, since they understood that it was essential
to restrict this menace if they were to sustain the needs of their
disparate empire.

The intervening period of half a millennium between Tyre
and Rome is seen also in the greater range of goods now traded.
These include pearls (probably from the Persian Gulf), silk (from
China!), as well as marble and specifically named spices such
as 'cinnamon, spice, incense, myrrh, frankincense' (Revelation
18.13). As expected, also listed are important staples to feed the
Empire, but besides the wine, oil and grain, there are now live

animals, cattle and sheep. Instruments or spoils of war are listed, too: horses and chariots, slaves – and human lives. Nonetheless, the splendour and desirability of such merchandise is captured here as effectively as in Ezekiel.

Despite the differences from Tyre's earlier cargo, many of the same criticisms apply equally to Rome: 'she glorified herself and lived luxuriously … In her heart she says, "I rule as a queen; I am no widow, and I will never see grief"' (v. 7). In other words, the same temptations of pride and arrogance and the tendency to assume that she is invincible apply to this later power as to the earlier one. However, there now appears to be horror at luxury of itself, though clearly the sin of 'Babylon' is perceived to go deeper than this as well. But there is something innately wicked about the longing of the soul for such possessions and in an excessive delight in wealth. The suddenness of her destruction is here expressed more starkly: the city that had (over time) 'grown rich' is destroyed in one hour (v. 19); and all her wealth, adorning her in the highest luxury, is laid waste in the same instant (v. 17). This powerfully – shockingly, even – demonstrates the transience of human efforts at self-aggrandisement, and their weakness before the divine wrath.

Accusations of violence feature here too, but there is a more general aroma of sin and repugnance for her deeds, unspecified as many of them may be: 'for her sins are heaped high as heaven, and God has remembered her iniquities' (v. 5).

This needs to be understood in the light of Rome being represented now as the embodiment of evil, for above all, as the final verse reminds us: '… in you was found the blood of prophets and of saints, and of all who have been slaughtered on earth' (v. 24).

Nonetheless, the characterization of 'Babylon' as mercantile, epitomizing greed and luxury, is a revealing line to take on Rome, given the extent of this great power's activities and (in the eyes of the writer of Revelation) multifarious sins. The corrupting aspect of trade on the grandest scale is clearly felt very deeply here.

Key message

The sea is and has long been a place of trade and interchange not just of material goods, but also of people and ideas. It is easy to interpret the biblical horror at 'Babylon's' (Rome's) excessive delight in her luxuries in terms of the massive gap in the ancient world between the rich and the poor. However, that gap is still present with us globally, in the gulf between the affluent 'First World' and the rest. In addition, the increasing trend towards the concentration of huge wealth in the hands of the few is well documented, yet this leaves the majority much less well off, and creates an ever-widening gap even between those on upper and middle incomes.[19] Furthermore, in the modern world, global trade entails the exploitation of the poor, as the Fairtrade movement makes clear. To take one example, textile workers in Bangladesh who work only for the export market receive only about a quarter of the wages of those in equivalent jobs serving the country's internal needs.[20] Worse still, global trade is often implicated in violence.[21] Both in biblical times and today, excessive wealth and economic dominance can appear both attractive and horrifying – but it needs to be limited and, despite the apparent hegemony of the super-rich, these biblical passages suggest that ultimately it can only lead to destruction.

Challenge

The economic dominance and excesses of the Western world can be both attractive, drawing those wishing to share in its prosperity, and repellent to those who question its underlying capitalist values. If we are to think of global markets and global economic systems, what level of average consumption is sustainable, and how can Western expectations be reduced? International interchange also prompts the exchange of ideas and culture. Should the dominant world culture (and economic ambitions) be permitted to become that of the materialistic, secular, consumerist, market-driven ethos of the West?

Economic activity on an international scale also has international implications, impacting on the weaker partner. Analysis of global supply chains reveals the extent to which millions of workers globally labour under poor living conditions to maintain the lifestyles of the affluent members of other societies. Global social inequalities are exemplified by a short supply-chain, in which tobacco is harvested in Tanzania (average income US$170), manufactured in Poland (average income US$10,000), and sold in France (average income US$58,000). As a result, even countries with high internal social equality can rely on labour from countries with high inequality.[22]

The environmental impact of certain practices which are motivated by wealth-generation or a desire for luxury products (such as tropical fruits or winter strawberries) can have similarly wide global effects. This is a further impetus to attempting to achieve sustainability. Even the more universal problems like global warming and rising sea levels, though largely fired by Western over-consumption, have a greater impact on poorer communities. For example, climate change, and in particular climate irregularity, can have a devastating effect on communities dependent on seasonal rains or living in regions where existence is already precarious. Compared to many of the world's greatest coastal cities (like London, New York, Tokyo, Sydney, Hong Kong, San Francisco, Cape Town and Shanghai), per capita expenditure on protection schemes is low for coastal cities in poorer areas of the world (such as in Addis Ababa, Lagos, Jakarta or Mumbai), even though rapid population growth and poorer-quality housing for many residents makes the need for investment all the more urgent.[23]

As many have noted previously, God seems to have a special concern for the poor and needy – the refugee, the stranger, the widow, the orphan – and so should we (Deuteronomy 14.29, 24.19–22; Psalm 146.9; James 1.27, 2.14–16). This has been recently emphasized in the Pope's latest encyclical, *Laudato Si'*, in which the extent to which ecological issues and economic injustice are inextricably interconnected is taken as a basis for much of what is said.[24]

Discussion and reflection

About ourselves

To what degree is our thinking and lifestyle dominated by the materialistic, secular, consumerist, market-driven ethos of the West? How would we answer the following questions:

- How rich do I consider myself? Is my perception correct?[25]
- How do I use the resources that I have? Perhaps discussing how we spend our money with others might be revealing (and challenging).
- How generous am I towards others who are less well off than myself?[26]
- How might I better use the resources that I have?

About others

Attitudes to migrants seeking refuge in Europe have sometimes been less than sympathetic, and in some quarters have focused on the minority who are economic migrants, or on a sense of being threatened by the values and religion that these people might bring, as well as on our own financial self-interest. Read the following poem by J. J. Bola, whose family had to flee the Democratic Republic of the Congo as refugees before settling in London when he was six years old:

'tell them (they have names)'

and when they turn the bodies over
to count the number of closed eyes. and they tell you
800'000: you say no. that was my uncle. he wore bright
coloured shirts and pointy shoes. 2 million: you say no.
that was my aunty. her laughter could sweep you up like
the wind to leaves on the ground.
6 million: you say no.
that was my mother. her arms. the only place i have ever
not known fear. 3 million: you say no. that was my love.
we used to dance. oh, how we used to dance.

or 147: you say no. that was our hope. our future.
the brains of the family.

and when they tell you that you come from war: you say no.
i come from hands held in prayer before we eat together.
when they tell you that you come from conflict: you say no.
i come from sweat.
on skin. glistening. from shining sun.
when they tell you that you come from genocide: you say no.
i come from the first smile of a new born child. gentle hands.
when they tell you that you come from rape: you say no.
and you tell them about every time you have ever loved.

tell them that you are from mother carrying you on her back.
until you could walk. until you could run. until you could fly.
tell them that you are from father holding you up to the night sky.
full of stars. and saying look, child. this is what you are made of.
from long summers. full moons. flowing rivers. and sand dunes.
tell them that you are an ocean that no cup could ever hold.

The poem strongly communicates the personal and human side of such tragedies. It has not just happened to 'them': it could be 'us'. The numbers are not just numbers: they are people, valued individuals, loved and lost by grieving families. There are many other compositions stemming from the experience of migration that convey the human cost and desperation of fleeing one's homeland or, as the Somali writer Warsan Shire expresses it in her poem 'Home' (reproduced on numerous websites),

no one leaves home unless
home is the mouth of a shark ...
you have to understand,
that no one puts their children in a boat
unless the water is safer than the land ...

The final words of this poem encapsulate the dreadful predicament that motivates the risks endured by fleeing refugees to reach safety: '... I know that anywhere is safer than here'. Having read these reflections on the experience of involuntary migration, now imagine what it might be like to be in the situation of fleeing your homeland for your own safety, or to suffer huge tragic losses within your family and community. What kind of response does this invite in us?

Migrants have always brought new ideas and skills, as well as different cultural practices. In fact, a study of White Europeans' DNA suggests that 80 per cent of us have remote ancestors who were ancient Near Eastern farmers (that is, from what is now the modern Middle East, comprising present-day countries such as Syria, Iraq and Iran) and who brought their newly developed agricultural skills with them in the Neolithic period to the hunter-gatherers of these islands.[27] The Neolithic agricultural revolution was one of the greatest turning points in the history of humanity, and even now many ideas, cultural practices, values and customs that migrant groups bring may have obvious potential to enrich our lives (for example, our cuisine). However, they can also present a challenge to our own expectations or traditional ways of doing things (for example, when a British mosque broadcasts a call to prayer, affecting the whole neighbourhood, or when the wearing of a burka may threaten our sense that reading facial signals is a prerequisite for open discourse), at times even giving rise to conflict.

Questions to consider arising from these issues might be:

- What benefits have immigrant communities brought in the past? And what problems may arise when different cultures try to coexist? How might these be resolved?
- How should we respond as Christians to the dispossessed? And how can we maintain a Christian identity and Christian values in a country in which the majority may no longer claim a Christian faith, and in which the 'no faith' and Muslim populations in particular are growing?

Action

We need to appropriate the idea that material wealth can be excessive, and to take active steps to reduce our ambitions and consumption. This might mean buying fairly traded or locally produced products, less meat, and more seasonal foods. But more than anything, it means buying less of everything. We need to change our attitude to consumption. The biblical passages we have looked at exemplify well the attractions of wealth and the status and power it can bring. Yet we need somehow to turn away from that mindset, and resist the compulsions of having and getting and spending. We should ask before making any purchase, 'Do I really need this? Or do I just want it?' This means taking seriously the reality that the earth is a single limited resource, and that when we have used its resources, there will be nowhere else to turn. What we have is a gift to be held in trust for those who come after us, not a vehicle for our own enrichment. For many, although this may require a serious rethink of priorities in life, it can also offer the liberating possibility of stepping off the treadmill of competitive consumption and allow a new way of living, freed from so many material concerns.

But as well as placing limits on our own acquisitiveness, alongside this we need to help those less well off than ourselves across the earth by generously supporting organizations such as Tearfund, Oxfam, Save the Children, and many others. In the West we are in an extraordinarily privileged position both in geographical and historical terms. Although we have so much that we are damaging the source of our prosperity, the earth itself, we can at least do something to help those who have less than the minimum they need even to live.

Notes

1 These verses are discussed in detail later in the chapter.
2 See L. Paine, 2014, *The Sea and Civilization: A Maritime History of the World*, London: Atlantic Books.
3 Indeed, with the satellite Global Positioning System (GPS) that we

are familiar with from the satnav in our cars, mariners can know with pinpoint accuracy where they are on the face of the earth and relative to land.

4 United Nations Environment Programme (UNEP) report.

5 See, for example, the WWF publication 'Principles for a Sustainable Blue Economy' www.wwf.se/source.php/1605623/15_1471_blue_economy_6_pages_final.pdf and M. Visbeck *et al.*, 2014, 'A Sustainable Development Goal for the ocean and coasts: global ocean challenges benefit from regional initiatives supporting globally coordinated solutions', *Marine Policy*, 49, pp. 87–9, in the context of UN Sustainable Development Goals.

6 This is written at the time immediately following the UK 'Brexit' referendum on EU membership, when the issues raised by migrants and refugees from around the world arriving in mainland Europe and the UK appear to be in the forefront of many people's concerns. This seems to be leading to a drawbridge mentality among some sections of UK society with respect to the English Channel.

7 See, for example, Deuteronomy 11.24, Joshua 1.4, Psalm 72.8, Micah 7.12 and Zechariah 9.10. For a variation on this, see Exodus 23.31, in which the Mediterranean and Euphrates still feature.

8 See, for example, Numbers 34.5–7, Deuteronomy 34.2, Zechariah 14.8, and also often in Joshua (for example, 15.12, 16.3, 6, 8, 17.9, 10, 19.29, 23.4) and Ezekiel 47—48 (47.15, 17, 19, 20, 48.28).

9 See www.freightafloat.co.uk/casestudies.asp?subarea=case%20studies&mi= for some case-studies of contexts in which freight is being reintroduced to the canals, to considerable environmental benefit.

10 Where 'Ophir' is mentioned in the Bible, it is always as a source of fine gold (1 Kings 9.28, 10.11, 22.48; 1 Chronicles 29.4; 2 Chronicles 8.18, 9.10; Job 22.24, 28.16; Psalm 45.9; Isaiah 13.12). Sheba is also associated with gold (Isaiah 60.5; Ezekiel 27.22), frankincense (Isaiah 60.5; Jeremiah 6.20), precious stones (Ezekiel 27.22) and spices (Ezekiel 27.22). The production of frankincense was confined to Arabia and (to a lesser extent) east Africa.

11 The translation of some of these terms is debated, as they are so rare. For example, the word here translated 'peacocks' may denote a type of monkey.

12 1 Kings 11.26–40, 12.1–24, and 2 Chronicles 10. The southern kingdom of Judah comprised the tribal areas of Judah and Benjamin, including the city of Jerusalem, while the much larger northern kingdom of Israel incorporated the remaining ten tribes.

13 Outside the purview of the books of Kings, but still within the scope of the Old Testament period, were the domination of the Persian and Greek empires. The experience of the first of these in particular could still have coloured the perspective of the writers or editors who helped 1 Kings reach its final form.

14 Note that the Tyrians were Phoenicians, seafarers who traded and established settlements across the Mediterranean, such as at Carthage in North Africa.

15 Besides places local to Syria-Palestine (vv. 5-6a, 17-18) and Trans-jordan (v. 16), as well as other Phoenician cities (vv. 8-9, and Arvad, v. 11), the list includes (according to our best efforts at identifying these places) Asia Minor (vv. 13-14) to the north-west, including Lydia (Lud, v. 10). To the west are Rhodes (v. 15), Cyprus (v. 6, and Elishah, which was probably a city on Cyprus, v. 7) and Tarshish (v. 12). Tarshish is an archetypally distant place, and its location is disputed, but it may even be in Spain. To the south are named Egypt (v. 7) and various locations in Arabia (vv. 20-22), as well as Put (in Africa, probably Libya or Somalia, v. 10). To the east are Assyria (Asshur, v. 23, including various Mesopotamian locations: Haran, Canneh and Eden, v. 23) and Persia (Paras, v. 10). Helech and Gamad (v. 11), Vedan and Uzal (v. 19) and Chilmad (v. 23) are all obscure. All this points to trade predominantly in the eastern Mediterranean, including both its northern and southern shores, but with some possible ventures further west (as is borne out by archaeological evidence and perhaps also by the mention of Tarshish). In addition, there was interchange across the Fertile Crescent (probably overland), but navigation down the Red Sea and possibly around Arabia may also be implied, though again some of this trade may have taken place overland.

16 Note that the attribute of being the 'perfection of beauty' should properly be applied to the ideal Jerusalem, as in Lamentations 2.15, where this city is also called 'the joy of all the earth'.

17 It has often been suggested that an English equivalent might be 'human from the humus'.

18 Similar language can also be identified in Psalms 18.4-19, 32.6, 69.13-15, 88.6-7 and 124.1-5.

19 On the widening gap not just between rich and poor, but between the rich and those who have so-called middle incomes, see Thomas Piketty, 2014, *Capital in the Twenty-First Century*, Cambridge, MA: Harvard University Press, and the recent report from Oxfam, *An Economy for the 1%: How Privilege and Power in the Economy Drive Extreme Inequality and How This Can Be Stopped* (available from http://policy-practice.oxfam. org.uk/publications/an-economy-for-the-1-how-privilege-and-power-in-the-economy-drive-extreme-inequ-592643). Essentially, the rich are getting richer and everyone else is getting poorer, to the extent that the world's richest 85 individuals collectively command as much wealth as the bottom 50 per cent of the global population. As a result, almost half of the world's wealth is owned by the richest 1 per cent, with the rest being divided among the remaining 99 per cent. It is worth noting that most of us in the Western world, if we have a job and income, are among the richest people on the earth! You can see how relatively rich you are by using one

of the online global wealth calculators – for example, www.givingwhatwe can.org/get-involved/how-rich-am-i/ (accessed 1 March 2016).

20 See Ali Alsamawi, Joy Murray and Manfred Lenzen, 2014, 'The employment footprints of nations: uncovering master–servant relationships', *Journal of Industrial Ecology*, 18.1, pp. 59–70.

21 On the violence associated with global trade, see Deborah Cowen, 2014, *The Deadly Life of Logistics: Mapping Violence in Global Trade*, Minneapolis, MN: University of Minnesota Press. Cowen highlights how preoccupation with supply-chain security has eroded workers' rights and internationalized employment practices. For example, Chinese ownership of the Greek port of Piraeus has entailed the importation of a Chinese labour model to Greece, with lowering of industry standards on a range of measures from training to working hours and wages, as well as a refusal to give recognition to unions. Similarly, the success of Dubai's Logistics City has led to the adoption of this model elsewhere, including in the USA, entailing close monitoring of workers and erosion of their rights. The identification of the free flow of goods as essential to national security has also led to a situation in which, according to Cowen, 'the maritime border becomes a space of transition: a zone subject to specialized government' (p. 83), 'where normal civil and labor law can be suspended' (p. 89). This has resulted in reliance on the transnationalization and privatization of security, and on the perception that disruption (for example, through industrial action) almost amounts to terrorism.

A further aspect of the conduct of global trade is the transference of security concerns effectively on to goods rather than people. Cowen also highlights the culpability and imperialism of the West in dealing with the problem of 'Somali piracy' in the Gulf of Aden, where Somali territorial waters have become governed and policed by private security and naval forces from various countries. Particularly concerning is the fact that the 'pirates' describe themselves as a volunteer coastguard, acting in response to the illegal dumping of toxic waste and illegal overfishing to the great detriment of local subsistence fishers by (in some cases) the very countries who have appointed themselves to act against the pirates. The essence of these complaints has been substantiated, which means the financial savings to those illegally dumping waste are almost certainly greater than the money extracted through piracy. The situation is complex: some of the culprits would say that waste normally disposed of at ports could not be deposited in this way because of the security concerns in Somalia.

22 See Joy Murray and Manfred Lenzen (eds), 2010, *The Sustainability Practitioner's Guide to Multi-Regional Input–Output Analysis*, Champaign, IL: Common Ground Publishing.

23 See Lucien Georgeson *et al.*, 2016, 'Adaptation responses to climate change differ between global megacities', *Nature Climate Change*, 6, pp. 584–8; or, more briefly, www.citylab.com/cityfixer/2016/03/the-climate-

change-spending-gap-adapt/472077/. For a specific example, see the following article on Bangladesh, highlighting how high population density, a weak infrastructure and low adaptive capacity have made the urban populations of this low-lying country especially vulnerable to climate change: Shamsuddin Shahid *et al.*, 2016, 'Climate variability and changes in the major cities of Bangladesh: observations, possible impacts and adaptation', *Regional Environmental Change*, 16.2, pp. 459–71.

24 Pope Francis, 2015, *Laudato Si': On Care for Our Common Home*, London: Catholic Truth Society.

25 Try using the global wealth calculator mentioned in note 19 above.

26 The Bible encourages us to be generous – for example, Deuteronomy 15.10–11, Psalm 112.5, Proverbs 11.25, and 1 Timothy 6.18.

27 Patricia Balaresque *et al.*, 2010, 'A predominantly neolithic origin for European paternal lineages', *PLoS Biology*, 8.1, e1000285, doi: 10.1371/journal.pbio.1000285.

9

Blue Planet, Blue God

... for the earth will be full of the knowledge of [the glory of] the LORD
as the waters cover the sea. (Isaiah 11.9; see also Habakkuk 2.14)

Having considered many aspects of the sea in the Bible we return
to the verse with which we began. What does it mean for the
waters to cover the sea? The context in Isaiah is the passage
(11.1–9) about the future rule of a righteous king and the coming
of a peaceable kingdom where:

> 6The wolf shall live with the lamb,
> the leopard shall lie down with the kid,
> the calf and the lion and the fatling together,
> and a little child shall lead them.
> 7The cow and the bear shall graze,
> their young shall lie down together;
> and the lion shall eat straw like the ox.
> 8The nursing child shall play over the hole of the asp,
> and the weaned child shall put its hand on the adder's den.
> 9They will not hurt or destroy
> on all my holy mountain;
> for the earth will be full of the knowledge of the LORD
> as the waters cover the sea. (Isaiah 11.6–9)

Here, 'as the waters cover the sea' simply means that the water
will fill the sea (basin) to capacity – that is, to fullness. God will be
present everywhere and will be fully known by all creation. Such a
beautiful poetic vision for peace and harmony, wisdom and security in the world speaks a powerful message. Within the Bible as a
whole it can be read in the light of the hope for Jesus' return and
the establishment of his reign on earth, which is understood to
include the promise of Isaiah 11. However, the New Testament

calls its readers not just to hope for such a kingdom and to look forward to Jesus' return when his reign will be established, but to seek to live in the present in the light of that future (eschatological)[1] hope.[2] It offers a vision of how things might be, inviting us to long and pray for its fulfilment, but also to seek to realize it, however imperfectly, now.

Self-evidently, we are not yet at the place where we see that fullness of the knowledge of God in creation, nor the transformation of creation into all that it is intended to be. Collectively living in pursuit of self-fulfilment rather than in hope for the wellbeing of all creation has, rather, made this vision seem more remote than it has ever been and caused damage unimagined by our biblical predecessors. This places especial responsibility on us to consider our response in the light of what we have studied in the previous chapters.

Blue Planet

As we noted at the beginning of the book, the oceans cover 71 per cent of the earth and our planet could equally well be called 'Ocean' as opposed to 'Earth'. Seen from space, our planet is indeed a beautiful 'blue marble' spinning in the vastness of the cosmos and, as far as we know, the only place in the universe with intelligent beings who can contemplate and understand something of themselves and the creation in which they live.[3]

The oceans may well be where life originated on our planet and they harbour life on all scales, from the smallest bacteria to the largest living creature, the blue whale. They also exhibit amazing biological diversity, perhaps best exemplified in the riot of colour and activity seen on coral reefs. The oceans are a source of food, carbon-based energy (in the form of undersea oil and gas) and renewable energy (wind, waves and tides). They comprise a barrier separating nations and peoples, but also a link enabling trade and travel between nations and peoples. They play a key part in the global economy, as evidenced by the use of terminology such as 'the blue economy', 'blue wealth' and 'blue growth'.

The oceans are also a source of pleasure and enjoyment. We

love sitting on a beach watching the waves, or swimming in the sea, or surfing or sailing on it. Snorkelling on a coral reef and watching the diversity of life there is an experience to be treasured by those who have had the opportunity to do this. For the more adventurous, diving deep beneath the waves is another source of delight in the ocean. Whale-watching from a boat at sea or seeing dolphins swimming in the water, or even swimming with them, allows us to take pleasure in the beauty of God's creation.

Although there are many positive aspects of the oceans, there are also some more terrifying facets. Being caught in a major storm in a small boat, as experienced by Jesus' disciples and by John Wesley and John Newton, can lead to fear for one's life. Tsunamis affect some parts of the world, bringing death, devastation and disaster to many people. This is a part of creation that is clearly beyond our human ability to control.

Though we cannot control the oceans we can seriously affect them in less than positive ways. This has been the case, particularly over the last century or so, as the planet's population has risen and our demands on the earth in terms of food and fuel have correspondingly increased. Humans have overfished the oceans leading to precipitous declines in certain fish stocks. We have allowed our refuse, particularly plastics and microplastics, to accumulate in the ocean, negatively affecting the marine life that ingests this. Our demand for energy, leading to more use of fossil fuels (oil, gas, coal), has pumped carbon dioxide into the atmosphere, warming the earth. Most of the extra heat (93 per cent) has gone into the ocean, thereby warming it. Likewise, the extra atmospheric carbon dioxide has been absorbed into the ocean, leading to ocean acidification. Both the warming and the acidification affect marine life, such as coral reefs, as well as the distribution of species across the global ocean (with cold-water species having to migrate towards the colder waters nearer the poles, away from the warming waters near the equator). There is as yet no indication that our seemingly insatiable demands on planet earth will abate in the short term, so there will be a continuing negative human impact on the oceans for the foreseeable future.

The earth is blue in colour – a beautiful blue planet, when seen from space – a perspective from which human degradation of the oceans is less obvious perhaps.[4] However, the planet may also be blue in the sense that it is sad over the state it is in. We are not here ascribing some goddess-like or pantheistic property to the earth, such as is done in some versions of the Gaia theory.[5] Rather, we are acknowledging that the Bible speaks of the earth mourning and suffering (Jeremiah 4.22–28; Hosea 4.1–3; cf. Isaiah 24.4–6). In both contexts this is the result of human sin. In our present-day context human greed and overexploitation of the earth and its oceans (sin by any other name) also leads to a metaphorically blue (sad and mourning) planet. What insights, then, does the Bible give as to what might be God's perspective on all this?

Blue God

God's perspective

Our exploration of the Bible has revealed that a key aspect of God's perspective on the oceans is his delight in his creation apart from any role we as humans may have in it. It has intrinsic value to him and was not created by him solely for the benefit of humanity. This is seen in many places in the Bible, starting in Genesis 1.20–21 where he creates the seas (cf. Psalm 95.5) and where he lets the waters teem with living creatures and sees that this is good prior to humans appearing on the scene. Psalm 104.24–26 similarly expresses God's creative power, and even his playfulness, in forming the great sea creature Leviathan to frolic in the ocean – an apparently pointless activity! We also noted God's response to Job in chapters 38 and 39 of that book, underscoring God's creative activity, which lies beyond our human understanding.

Another key theme that has emerged is God's command of and power over the oceans and its winds and waves. This is seen in both the Old Testament (for example, Job 9.8, Psalms 65.7, 93.3–4, 107.24–25) and in the New, where Jesus commands the winds and waves (Mark 4.35–41) and walks on the water (Mark 6.47–51). What may appear terrifying to us is not so to God, as it is part of his creation, over which he rules, and which obeys his commands.

Ultimately, as Psalm 24.1 states, 'the earth is the LORD's, and all that is in it.' A biblical perspective is that the earth is his, but he has entrusted it to humanity (Psalm 115.16; cf. Genesis 2.15, 1.26, 28), though he still has power to command creation and it will obey him. Sadly, we have failed to look after what he entrusted to us, with consequences not just for ourselves – but also for other creatures and for the earth itself.

God's pathos

No doubt our failure to care for the creation grieves him and causes him pain. It is fair, then, to say that God is 'blue' – that is, he is sad and disappointed in how his children have sinned and are in the process of damaging his good creation. As we look at the Bible, we can see many occasions when humanity generally, and God's people in particular, managed to grieve God. For example, a precursor to the flood narrative in Genesis was that God was grieved by humanity's wickedness and was sorry to have made them on the earth (Genesis 6.5–6). The consequences both for humanity and other creatures were severe, again emphasizing that human sin impacts the rest of creation negatively (cf. Hosea 4.1–3). Israel too grieved God by their actions after he rescued them from Egypt in the Exodus (Psalm 78.40). The author of the letter to the Ephesians warns his readers not to 'grieve the Holy Spirit of God' by the way they live (Ephesians 4.30), and that call echoes down to our own time. Fortunately, that is not the end of the story.

God's patience

The Bible offers three main messages in response to human wrongdoing: judgement, divine forbearance, and hope. It seeks to hold together two seemingly opposing characteristics of God: he is a compassionate and loving Father (Exodus 34.6; Psalm 145.8–9; Matthew 7.9–11) as well as a God of righteousness and justice (Psalm 11.7, 33.5). This was a lesson Jonah found hard to learn despite his experience of God at sea because when he delivered his message of judgement to the people of Nineveh and they repented and were forgiven, we are told:

¹But this was very displeasing to Jonah, and he became angry. ²He prayed to the LORD and said, 'O LORD! Is not this what I said while I was still in my own country? That is why I fled to Tarshish at the beginning; for I knew that you are a gracious God and merciful, slow to anger, and abounding in steadfast love, and ready to relent from punishing. ³And now, O LORD, please take my life from me, for it is better for me to die than to live.' (Jonah 4.1-3)

In the modern world, we are more inclined the other way, and are likely to resist the model of God that Jonah was advocating, instead trusting that he should be 'gracious ... and merciful, slow to anger, and abounding in steadfast love, and ready to relent from punishing' without limit. However, the Bible does not just depict judgement as coming from 'on high', imposed by God as a punishment. Consistently, we are shown a picture of actions having consequences, and human wrongdoing often having an impact on the health of the environment in which (and sometimes beyond which) the perpetrators lived. This was not an understanding that was derived scientifically, but grew out of a simple appreciation of the interconnectedness of the wellbeing of all of us, of goodness bringing blessing, and of sin being an unhealthy state that was bound to have negative consequences. The Bible also offers a passionate advocacy for justice, that if the weak are exploited or corruption affects society, this is something that must be put right. Not suffering the consequences of our actions prevents us from learning and growing, but also seriously undermines principles of justice – as indeed any parent will immediately appreciate.

Ultimately, the Bible, through the New Testament, offers a vision of hope that incorporates both judgement and a new and better order. The Epistle to the Romans was written in dark times of persecution of members of the new Christian community in Rome. Mirroring this situation, Paul speaks of how 'the whole creation has been groaning in labour pains until now' (Romans 8.22), suffering so that the new order can emerge. 'Creation waits with eager longing' (v. 19), he says, for 'the creation itself will be set free from its bondage to decay and will obtain the freedom of the glory of the children of God' (v. 21). The hope for all creation is taken up also in Colossians, which emphasizes that Jesus is the Creator of 'all things', and that his purpose of reconciliation was for all creation too:

> ... [16]in him all things in heaven and on earth were created ... all things have been created through him and for him. [17]He himself is before all things, and in him all things hold together ... [20]and through him God was pleased to reconcile to himself all things, whether on earth or in heaven, by making peace through the blood of his cross. (Colossians 1.16–20)

This backdrop of judgement and hope offers a perspective in which we can act, knowing both that we must take the consequences of our behaviour in full seriousness, but also holding fast to a vision of the future. Some actions are clearly incompatible with the end point that the Bible understands to be what God intends for both the earth and for his people.[6] His people are supposed to be a sign of what is to come here on earth.[7] What, then, should our response be to the challenges of living in a good, but damaged, creation?

Blue people

Taking into account both scientific and biblical perspectives on the ocean, how then are we to live? We have already noted that God's people are to reflect his character on the earth and to carry out his will. The previous chapters have used the Bible's perspectives on the sea as a lens through which we can examine how we live our lives. Looking back over the challenges posed in those chapters we can see that concern for the welfare of the sea and its creatures, for our fellow human beings, and for the planet as a whole, has to be paramount if we are to reflect God's concern for his whole creation and to respond to our current environmental crisis. This puts each of us in a position of responsibility (and indeed of culpability), since degradation of the planet or exploitation of individuals for commercial gain is the cumulative product of the decisions and choices made by each one of us. It is paramount for each of us to consume less, not just by chipping away at our energy use (though that can help), but by changing our relationship with 'stuff' and making thoughtful choices about what we really need based on what we can do without rather than what we might like to have if we lived in a limitless world.

At the same time, we also need to think more humbly about our own place in the world. As well as relinquishing inappropriate aspirations to power and control over our environment[8] and other species, we also need to recognize the vulnerability of the earth and the fragility of even the vast yet vulnerable sea. This should lead us to adopt a humbler and more appropriate attitude to our planet, as well as recognizing the precariousness not just of the natural systems on which we depend, but our own existence too.

This should also entail an acceptance of our responsibilities as the only species able to have such an immense impact on the entire planet. We alone understand something of what we are doing and should be able to co-ordinate collective action and a common sense of purpose to address the consequences of our own behaviour. This links in with the scientifically demonstrated reality of anthropogenic change, but also echoes (albeit from a different starting point) the biblical insight that human behaviour can affect the wellbeing of all creation. Moreover, if we acknowledge biblical claims to a unique place in creation for humanity, as made in the image of God and 'a little lower than God'[9] (as Psalm 8 puts it), this must necessitate a high level of responsibility to the rest of creation and indeed to God himself. If Israel discovered that election (being chosen by God) meant greater responsibility and therefore greater guilt and reason for judgement before God (Amos 3.2), what should become of a species that has so misused its God-given gifts and failed to 'do justice, and to love kindness, and to walk humbly with [our] God' (Micah 6.8)?

We need to live in harmony with God's purposes for his creation, mindful of the 'sacredness' of the sea, and seeking not to overstep the limits set for us. It also means recognizing that there is no neutral ground: not making the lifestyle and attitude changes required is an active decision, entailing responsibility (and, yes, guilt), not a passive one. By doing nothing, we are directly contributing to the ruin of God's good earth.

Becalmed?

It is a strange paradox that climate change and environmental degradation are almost universally acknowledged and yet we have, so far, collectively made very little progress in combating their effects. Like a sailing ship when there is no wind, we seem becalmed, unable to move or act. Much of this is due to very human factors: lack of prioritization, financial cost, conflicting ideologies, social and economic influences, lack of trust and risk aversion. We continue to drive our cars or keep the heating as it has always been set, or leave electrical goods on standby because this is what we are accustomed to doing. Changing deeply ingrained habits takes motivation and effort, because the default option is to follow well-established behaviour patterns without ever examining them critically.

We have evolved to deal primarily with present or short-term threats and with our immediate surroundings and social context. We can transcend that 'short-termism' to some extent, but responding to climate change, on the whole, requires the capacity to take seriously threats that may present themselves beyond our lifetimes. It is a major challenge to respond as if these problems were personally and immediately pressing, even though they are not so in any visible and tangible way. If we feel disinclined to take ecological and environmental issues seriously, we can buy newspapers, or listen to and watch news channels, or read material on the web, that reinforce a view that justifies our apathy. This is exacerbated by mistrust of declarations by politicians, journalists and even scientists – those with 'vested interests'. Can we really believe what is claimed?

Living in a more sustainable way may also require social courage. If we feel that we shall be judged by our car or our house or our clothes, it is tempting to buy what we do not need in order to conform to society's materialistic norms. Standing up against these is not just an act of resistance, but also a change that may liberate others to follow, and could in time reform some of the values of our society and create new norms. Indeed, a new cultural attitude and cultural values are essential for reform to take place. However, to lead the way and resist others' expectations also takes courage.

It can, moreover, be very difficult to reconcile competing convictions and to alter deeply held values and beliefs. If we have benefited financially and personally from the current economic system, if we enjoy the products and benefits of a consumerist culture and cannot see a viable alternative to economic growth and increased productivity except one that would lead to disaster, we are likely, in practice, not to respond at a fundamental level to the challenges of the current environmental crisis. If the status quo suits us, we may be resistant to change. And we can always hope for salvation through technological innovation, and for decision-making and action by governments, thus enabling us to continue to live as we always have done.

Beyond this, the fear of losing out, of making sacrifices while others do not, also inhibits real transformation. Why should I stand in the rain waiting for a bus while my neighbour drives into work? Why should I resist fishing for a particular species in a way that may cause irreversible decline if I know someone else will only take what I leave? Better to profit now while I can than to miss out altogether.

In addition, moderating our impact on the environment may involve personal sacrifice: it will entail decreased consumption, maybe adjusting to a less warm home or a smaller car. It may involve personal risk: cycling when the majority of people are still car-bound and not necessarily mindful of cyclists may be dangerous. And it is likely to entail financial cost, whether we choose to install solar panels or buy locally produced vegetables.

A *way through the sea*[10]

In the light of our previous discussion and our study of the Bible, how might we 'navigate' our way better through the complexities of life in the modern world? In each chapter we have tried to provide material for the reader to reflect on and suggestions for possible actions that could be taken in response. Our aim has not been to tell the reader what to do *per se*, but to provide insights from the Bible that might provoke us and unsettle our twenty-first century perspectives. We all need to be challenged as, being human, we tend to default to allowing inertia to keep us continuing on the same path through life as we have always followed – it is the easy option. Making a decision to change – and then actually carrying it out – takes a certain amount of resolve that many of us seem to lack much of the time. As we have seen earlier in this book, a dramatic experience of danger at sea can be a catalyst for such change.

Unfortunately, a book cannot provide that experience but hopefully it can make us aware that our often-comfortable lives are not the only way to live, nor necessarily the best way. The vastness of the ocean, and the forces beyond our control that are associated with it – such as storms and tsunamis – remind us of our puniness. That awareness should keep us humble and induce in us a degree of humility with regard to our interactions with the sea and the earth and the creatures in and on them. Do we exhibit such humility in our lives?

When faced with all the challenges of modern life it can be easy to despair as some of the problems, such as the environmental ones, may seem insurmountable. However, many diverse voices – not necessarily religious ones – see that faith can have an especially valuable role to play in this context.[11]

For readers who are not Christians, we hope that much of what we have outlined, both from a biblical and scientific standpoint, resonates as a perspective that needs to be taken seriously: that the ocean should be valued of itself, and that we must urgently reform our consumption-based outlook and live more justly and sustainably. Belief in the right course of action and a willingness to carry it through does not have to be based on religious premises, and nor do you have to be a committed member of a faith community in order to recognize the value of much of what the Bible says. Nonetheless, the effects of faith and the practice of prayer, and being part of a wider community of people living according to the same aims and values, can help.

If we truly believe that God created and values the oceans, we should make this our concern too. It may be that the way forward, like the sea at times, seems chaotic and our ability to make right choices uncertain. Nevertheless, like the apostle Peter (Matthew 14.22–33), it is perhaps better to step out of the boat and risk sinking than not step out of the boat at all. That is, if we have been challenged about how we should live through the perspectives offered by the Bible on the sea then we should act accordingly. What the future holds we cannot predict, but we should live life as if we are, as Wesley put it, at 'every moment on the brink of Eternity', mindful always of this ultimate perspective. This, if anything, should equip us to respond.

Setting sail

As we embarked on our exploration of biblical and scientific understandings of the sea, we began this book with a prayer of St Brendan the Navigator. Now we have reached the end of that particular journey, what we have discovered invites a new perspective and a new direction of travel. Perhaps, therefore, as we seek to reorient ourselves in order to live in the light of the biblical perspectives on the sea, it is appropriate that we set sail on the next stage of the journey with another of his prayers:

> Help me to journey beyond the familiar
>   and into the unknown.
> Give me the faith to leave old ways
>   and break fresh ground with You.
>
> Christ of the mysteries, I trust You
>   to be stronger than each storm within me.

I will trust in the darkness and know
   that my times, even now, are in Your hand.
Tune my spirit to the music of heaven,
   and somehow, make my obedience count for You.

Notes

1 Eschatology, from two Greek words meaning 'last' (*eschatos*) and 'talk' (*logos*), is the study of 'end things': the end of the age, the end of the world, and the nature of the kingdom of God.

2 On Christian hope, see Tom Wright, 2007, *Surprised by Hope*, London: SPCK.

3 On this topic, see David Wilkinson, 2013, *Science, Religion, and the Search for Extraterrestrial Intelligence*, Oxford: OUP.

4 Nonetheless, the shrinkage of the Aral Sea, Lake Chad, the Sea of Galilee and other inland lakes and seas due to human activity (see Chapter 4) has been clearly observed using satellite sensors flown in space. See www.ibtimes.co.uk/world-water-day-2016-before-after-images-earths-disappearing-lakes-seas-1550383, and the discussion on the Earth Policy Institute website (Janet Larsen, 'Disappearing Lakes, Shrinking Seas', under 'Plan B Updates').

5 The Gaia hypothesis or theory was formulated by James Lovelock (2016, *Gaia: A New Look at Life on Earth*, reprint, Oxford: OUP; original published in 1979) and posits that the earth is a self-regulating, complex system that operates to maintain the right conditions for life on earth (though not necessarily human life, it should be noted). Some have taken this scientific theory as support for a neo-pagan view of the earth, Gaia being the ancient Greek goddess of the earth, though Lovelock himself does not. For a critical scientific assessment of Gaia theory, see T. Tyrell, 2013, *On Gaia: A Critical Investigation of the Relationship between Life and Earth*, Princeton, NJ: Princeton University Press.

6 For more on this issue of living in the light of the future, see Wright's book mentioned in note 2 and also M. A. Srokosz, 2008, 'God's story and the earth's story: grounding our concern for the environment in the biblical metanarrative', *Science & Christian Belief*, 20, pp. 163–74.

7 John Howard Yoder expresses this in terms of an ideal picture of what the Church should be: 'Although immersed in this world, the church by her way of being represents the promise of another world, which is not somewhere else but is to come here' (John Howard Yoder, 1984, *The Priestly Kingdom: Social Ethics as Gospel*, Notre Dame, IN: University of Notre Dame Press). However, this should not be understood only to relate to the Church in a formal, institutional sense. It is a vision to which

anyone, as an individual, can seek to respond, even if it is in practice easier to do so in the company of a community of similarly minded people, seeking to live out the same vision.

8 Ironically, events like tsunamis and hurricanes soon disabuse us of any thought that we might be able to exercise power or control over our environment.

9 This expression occurs in a possibly more familiar form – 'a little lower than the angels'– in Hebrews 2.7, which quotes this psalm. This difference occurs because the Hebrew rendered into the New Testament as 'angels' can be translated either as 'God' or 'gods' (that is, the divine beings comprising members of his heavenly court). The NRSV translation, which we are following here, assumes the former (which is the more common use of the Hebrew Elohim), but the Septuagint (the Greek translation of the Hebrew Bible which was familiar to the writers of the New Testament) assumes the latter. The word 'angels', then, simply interprets and clarifies the reference to the divine beings so that the reader knows to read it within a monotheistic framework, and to recognize their subordinate status as compared with God himself. This in turn implies a lesser, though still exalted, status for humanity: not a little lower than God, but than his angels (literally messengers).

10 From Isaiah 43.16:

This is what the LORD says –
he who made a way through the sea,
a path through the mighty waters.

11 See, for example, The Environment Agency's 2007 report '50 Things that will Save the Planet', which drew on the views of a panel of leading scientists and environmentalists. It judged the role that organized religion and faith leaders may play as of second most importance in terms of its potential to save the earth. Similarly, the 1992 World Scientists' *Warning to Humanity*, signed by about 1700 leading scientists including most of the Nobel Laureates in the sciences, enlisted the help of faith leaders (among others) in addressing the environmental crisis. More recently, an article in *Science* (November 2014) written by an economist and an oceanographer urged the need for religious institutions to mobilize public opinion and action (Partha Dasgupta and Veerabhadran Ramanathan, 2014, 'Pursuit of the common good', *Science*, 345/6203, pp. 1457–8).

Further Reading

Sea Poetry Anthologies

Crew, Bob, 2005, *Sea Poems: A Seafarer Anthology*, Rendlesham: Seafarer Books.

Jay, Peter (ed.), 2005, *The Sea! The Sea! An Anthology of Poems*, London: Anvil Press Poetry.

McClatchy, J. D., *Poems of the Sea*, 2001, Everyman's Library Pocket Poets; New York: Alfred A. Knopf.

Watson, Howard, 2011, *Ode to the Sea: Poems to Celebrate Britain's Maritime Heritage*, London: National Trust Books.

The Sea in Literature

Guite, Malcolm, 2017, *Mariner: A Voyage with Samuel Taylor Coleridge*, London: Hodder & Stoughton.

Jefferson, Sam, 2015, *Sea Fever: The True Adventures that Inspired our Greatest Maritime Authors, from Conrad to Masefield, Melville and Hemingway*, London: Adlard Coles Nautical.

Klein, Bernhard, 2002, *Fictions of the Sea: Critical Perspectives on the Ocean in British Literature and Culture*, Farnham: Ashgate Publishing.

Adventures at Sea: Journals, Biographies and Autobiographies

Aitken, Jonathan, 2007, *John Newton: From Disgrace to Amazing Grace*, London: Continuum.

Barron, W. R. J. and Burgess, G. S. (eds), 2005, *The Voyage of St Brendan: Representative Versions of the Legend in English Translation*, Exeter: University of Exeter Press.

Newton, John, 1765, *An Authentic Narrative of Some Remarkable and Interesting Particulars in the Life of Mr John Newton: Communicated in a Series of Letters to the Reverend Mr Haweis, Rector of Aldwinckle*,

Northamptonshire, and by him (at the Request of Friends) Now Made Public, 3rd edn, London: printed for S. Drapier, T. Hitch and P. Hill (accessible online).

MacArthur, Ellen, 2010, *Full Circle: My Life and Journey*, London: Michael Joseph.

Severin, T., 1978, *The Brendan Voyage: Across the Atlantic in a Leather Boat*, London: Hutchinson & Co.

Wesley, John, 1951, The Journal of John Wesley (Grand Rapids: Christian Classics Ethereal Library; Chicago, IL: Moody Press, 1951), available at http://www.ccel.org/ccel/wesley/journal.html.

Oceanography

Casey, Susan, 2011, *The Wave: In Pursuit of the Ocean's Deadliest Furies*, London: Vintage Books.

Denny, Mark, 2008, *How the Ocean Works: An Introduction to Oceanography*, Princeton, NJ: Princeton University Press.

Kunzig, Robert, 2010, *Mapping the Deep: The Extraordinary Story of Ocean Science*, London: Sort of Books.

Stow, Dorrik, 2005, *Encyclopedia of the Oceans*, Oxford: Oxford University Press.

Stow, Dorrik, 2017, *Oceans: A Very Short Introduction*, Oxford: Oxford University Press.

The Bible and the Environment

Baukham, Richard, 2010, *Bible and Ecology: Rediscovering the Community of Creation*, London: Darton, Longman & Todd.

Habel, Norman, 2009, *An Inconvenient Text: Is a Green Reading of the Bible Possible?*, Adelaide: ATF Press.

Horrell, David G., 2010, *The Bible and the Environment: Towards a Critical Ecological Biblical Theology*, London: Equinox.

Horrell, David G. *et al.* (eds), 2010, *Ecological Hermeneutics: Biblical, Historical and Theological Perspectives*, London: T&T Clark.

Theology and the Environment

Berry, R. J. (ed.), 2006, *Environmental Stewardship: Critical Perspectives – Past and Present*, London: T&T Clark.

Bookless, Dave, 2008, *Planetwise: Dare to Care for God's World*, Nottingham: Inter-Varsity Press.

Bouma-Prediger, 2010, *For the Beauty of the Earth: A Christian Vision for Creation Care*, 2nd edn, Grand Rapids, MI: Baker Academic.

Conradie, Ernst M., 2015, *The Earth in God's Economy: Creation, Salvation and Consummation in Ecological Perspective*, Studies in Religion and the Environment, 10; Vienna: LIT Verlag.

Pope Francis, 2015, *Encyclical Letter Laudato Si' of the Holy Father Francis: On Care for our Common Home*, London: Catholic Truth Society.

White, Robert S. (ed.), 2009, *Creation in Crisis: Christian Perspectives on Sustainability*, London: SPCK.

The Marine Environment

Patton, Kimberley C., 2007, *The Sea Can Wash Away All Evils: Modern Marine Pollution and the Ancient Cathartic Ocean*, New York: Columbia University Press.

Creation

Barker, Margaret, 2010, *Creation: A Biblical Vision for the Environment*, London: T&T Clark.

Doak, Brian R., 2014, *Consider Leviathan: Narratives of Nature and the Self in Job*, Minneapolis, MN: Fortress Press.

Simkins, Ronald A., 1994, *Creator and Creation: Nature in the Worldview of Ancient Israel*, Peabody, MA: Hendrickson Publishers.

The Origins of Life

Holder, Rodney D., 2013, *Big Bang, Big God: A Universe Designed for Life?*, Oxford: Lion Hudson.

Lane, Nick, 2016, *The Vital Question: Why is Life the Way it is?*, London: Profile Books.

The Temple

Barker, Margaret, 1991, *The Gate of Heaven: The History and Symbolism of the Temple in Jerusalem*, London: SPCK.

Water

Pearce, Fred, 2006, *When the Rivers Run Dry: What Happens When our Water Runs Out?*, London: Eden Project Books, Transworld Publishers.
Peppard, Christiana Z., 2014, *Just Water: Theology, Ethics, and the Global Water Crisis*, New York: Orbis Books.
Younger, Paul L., 2012, *Water: All that Matters*, London: Hodder & Stoughton.

Chaos

Gleick, James, 1997, *Chaos: Making a New Science*, London: Vintage.
Levenson, Jon D., 1994, *Creation and the Persistence of Evil: The Jewish Drama of Divine Omnipotence*, Princeton, NJ: Princeton University Press.
Smith, Lenny, 2007, *Chaos: A Very Short Introduction*, Oxford: Oxford University Press.
White, R. S., 2014, *Who is to Blame? Disasters, Nature and Acts of God*, Oxford: Monarch.
Wuthnow, Robert, 2010, *Be Very Afraid: The Cultural Response to Terror, Pandemics, Environmental Devastation, Nuclear Annihilation, and other Threats*, Oxford: Oxford University Press.

Ancient Trade and Travel

Abulafia, David, 2011, *The Great Sea: A Human History of the Mediterranean*, London: Penguin Books.
Casson, Lionel, 1994, *Ships and Seafaring in Ancient Times*: Texas: University of Texas Press.
Paine, L., 2014, *The Sea and Civilization: A Maritime History of the World*, London: Atlantic Books .
Patai, Raphael, 1998, *The Children of Noah: Jewish Seafaring in Ancient Times*, Princeton, NJ: Princeton University Press.

Modern Economy and Sustainability

Cowen, Deborah, 2014, *The Deadly Life of Logistics: Mapping Violence in Global Trade*, Minneapolis, MN: University of Minnesota Press.
Mitchell, Timothy, 2011, *Carbon Democracy: Political Power in an Age of Oil*, London: Verso.

Piketty, Thomas, 2014, *Capital in the Twenty-First Century*, Cambridge, MA: Harvard University Press.

Raworth, Kate, 2017, *Doughnut Economics: Seven Ways to Think Like a 21st-Century Economist*, London: Random House Business Books.

Green Economics

Cato, Molly Scott, 2009, *Green Economics: An Introduction to Theory, Policy and Practice*, Abingdon: Earthscan.

Cato, Molly Scott, 2011, *Environment and Economy*, Abingdon: Routledge.

Dietz, Rob, and O'Neill, Dan, 2013, *Enough is Enough: Building a Sustainable Economy in a World of Finite Resources*, Abingdon: Earthscan.

Jackson, Tim, 2009, *Prosperity without Growth: Economics for a Finite Planet*, Abingdon: Earthscan.

Klein, Naomi, 2015, *This Changes Everything: Capitalism v. the Climate*, London: Penguin.

The Bigger Biblical Picture: Metanarrative and the World to Come

Murray, Robert, 1992, *The Cosmic Covenant: Biblical Themes of Justice, Peace and the Integrity of Creation*, Heythrop Monographs 7; London: Sheed & Ward.

Wright, Christopher, 2006, *The Mission of God: Unlocking the Bible's Grand Narrative*, Nottingham: Inter-Varsity Press.

Wright, Christopher, 2010, *The Mission of God's People*, Grand Rapids, MI: Zondervan.

Wright, Tom, 2007, *Surprised by Hope*, London: SPCK.

Science and Religion

Southgate, Christopher (ed.), 2011, *God, Humanity and the Cosmos: A Textbook in Science and Religion*, 3rd edn, London: T&T Clark.

Wilkinson, David, 2013, *Science, Religion, and the Search for Extraterrestrial Intelligence*, Oxford: Oxford University Press.

Biblical Ethics

Barton, John, 2002, *Ethics and the Old Testament*, 2nd edn, London: SCM Press.

Rodd, Cyril, 2001, *Glimpses of a Strange Land: Studies in Old Testament Ethics*, Edinburgh: T&T Clark.

Wright, Christopher J. H., 2004, *Old Testament Ethics for the People of God*, Leicester: Inter-Varsity Press.

Yoder, John Howard, 1984, *The Priestly Kingdom: Social Ethics as Gospel*, Notre Dame, IN: University of Notre Dame Press.

Index of Bible References

Index of Names and Subjects